Supplements to the 2nd Edition of

RODD'S CHEMISTRY OF CARBON COMPOUNDS

ELSEVIER SCIENCE PUBLISHERS B.V.
Sara Burgerhartstraat 25
P.O. Box 211, 1000 AE Amsterdam, The Netherlands

Distributors for the United States and Canada:

ELSEVIER SCIENCE PUBLISHING COMPANY INC.
52, Vanderbilt Avenue
New York, N Y 10017

Library of Congress Card Number: 64-4605

ISBN 0-444-42821-6

Printed in The Netherlands

Supplements to the 2nd Edition of

RODD'S CHEMISTRY OF CARBON COMPOUNDS

VOLUME I

ALIPHATIC COMPOUNDS
★

VOLUME II

ALICYCLIC COMPOUNDS
★

VOLUME III

AROMATIC COMPOUNDS
★

VOLUME IV

HETEROCYCLIC COMPOUNDS
★

VOLUME V

MISCELLANEOUS
GENERAL INDEX
★

6a

Supplements to the 2nd Edition (Editor S. Coffey) of

RODD'S CHEMISTRY OF CARBON COMPOUNDS

A modern comprehensive treatise

Edited by
MARTIN F. ANSELL
Ph.D., D.Sc. (London) F.R.S.C. C. Chem.
Reader Emeritus, Department of Chemistry,
Queen Mary College, University of London, Great Britain

Supplement to

VOLUME IV HETEROCYCLIC COMPOUNDS

Part F:
Six-Membered Heterocyclic Compounds with a Single Nitrogen Atom in the Ring; Pyridine, Polymethylenepyridines, Quinoline, Isoquinoline and Their Derivatives

ELSEVIER
Amsterdam — Oxford — New York — Tokyo   1987

CONTRIBUTORS TO THIS VOLUME

JAN BECHER
Department of Chemistry, Odense University,
DK 5230 Odense, Denmark

C. DAVID JOHNSON
School of Chemical Sciences, University of
East Anglia, Norwich, NR4 7TJ

JAMES G. KEAY
Reilly Tar and Chemical Corporation, 151 North Delaware
Street, Indianapolis, Indiana 46204, U.S.A.

ERIC F. V. SCRIVEN
Reilly Tar and Chemical Corporation, 151 North Delaware
Street, Indianapolis, Indiana 46204, U.S.A.

RAYMOND E. FAIRBAIRN
Formerly of Research Department,
Dyestuffs Division,
I.C.I. (INDEX)

PREFACE TO SUPPLEMENT IVF

The publication of this volume continues the supplementation of the second edition of Rodd's Chemistry of Carbon Compounds thus keeping this major work of reference up to date. This supplement covers the chapters in Volume IVF of the second edition with the exception of Chapter 23. The latter chapter in the second edition surveyed the theoretical chemistry and spectroscopic properties of heterocycles containing a pyridine ring. In this supplement these aspects of the classes of compounds reviewed have been incorporated in the relevant chapters and are not discussed separately. Although each chapter in this book stands on its own, it is intended that it should be read in conjunction with the parent chapter in the second edition.

As editor I wish to express my thanks to each of the contributors for all the effort and expertise they have put into producing their chapter. The areas of chemistry surveyed are important ones with a very extensive literature. In each case the author has produced a very readable, concise and critical account of the particular field of chemistry reviewed. I also wish to thank Dr Fairbairn for compiling the index.

At a time when there are many specialist reviews, monographs and reports available, there is still in my view an important place for a book such as "Rodd" which gives a broader coverage of organic chemistry. One aspect of the value of this work is that it allows the expert in one field to quickly find out what is happening in other fields of chemistry. It provides a convenient means into a field of study providing an outline of the chemistry in that area together with leading references to other works which provide more detailed information.

This volume has been produced by direct reproduction of the manuscripts. I am most grateful to the contributors for all the care and effort both they and their secretaries have put into the production of the manuscripts and diagrams. I also wish to thank the staff at Elsevier for all the help they have given me and for seeing the transformation of the authors' manuscripts to published work.

April 1987   Martin Ansell

CONTENTS

VOLUME IV F

Heterocyclic Compounds: Six-Membered Heterocyclic Compounds with a Single Nitrogen Atom in the Ring; Pyridine, Polymethylenepyridines, Quinoline, Isoquinoline and Their Derivatives

Chapter 24. The Chemistry of Pyridine and its Derivatives*
by J. BECHER

---

*The numbering of sections and sub-sections of Chapter 24 corresponds to the parent chapter in the second edition of Volume IV F. Consequently, some sub-sections not covered in this supplement are omitted, e.g. 3(b)(iv).

Chapter 25. Bicyclic Compounds Containing a Pyridine Ring:
Cyclopolymethylenepyridines, Cycloalkenopyridines
by C.D. JOHNSON

Chapter 26. Bicyclic Compounds Containing a Pyridine Ring; Quinoline and its
Derivatives
by C.D. JOHNSON

*Chapter 27. Bicyclic Compounds Containing a Pyridine; Isoquinoline and its
Derivatives*
by J.G. KEAY and E.F.V. SCRIVEN

OFFICIAL PUBLICATIONS

B.P.	British (United Kingdom) Patent
F.P.	French Patent
G.P.	German Patent
Sw.P.	Swiss Patent
U.S.P.	United States Patent
U.S.S.R.P.	Russian Patent
B.I.O.S.	British Intelligence Objectives Sub-Committee Reports
F.I.A.T.	Field Information Agency, Technical Reports of U.S. Group Control Council for Germany
B.S.	British Standards Specification
A.S.T.M.	American Society for Testing and Materials
A.P.I.	American Petroleum Institute Projects
C.I.	Colour Index Number of Dyestuffs and Pigments

SCIENTIFIC JOURNALS AND PERIODICALS

With few obvious and self-explanatory modifications the abbreviations used in references to journals and periodicals comprising the extensive literature on organic chemistry, are those used in the World List of Scientific Periodicals.

LIST OF COMMON ABBREVIATIONS AND
SYMBOLS USED

A	acid
Å	Ångström units
Ac	acetyl
a	axial; antarafacial
as, $asymm.$	asymmetrical
at	atmosphere
B	base
Bu	butyl
b.p.	boiling point
C, mC and μC	curie, millicurie and microcurie
c, C	concentration
C.D.	circular dichroism
conc.	concentrated
crit.	critical
D	Debye unit, 1 x 10^{-18} e.s.u.
D	dissociation energy
D	dextro-rotatory; dextro configuration
DL	optically inactive (externally compensated)
d	density
dec. or decomp.	with decomposition
deriv.	derivative
E	energy; extinction; electromeric effect; Entgegen (opposite) configuration
E1, E2	uni- and bi-molecular elimination mechanisms
E1cB	unimolecular elimination in conjugate base
e.s.r.	electron spin resonance
Et	ethyl
e	nuclear charge; equatorial
f	oscillator strength
f.p.	freezing point
G	free energy
g.l.c.	gas liquid chromatography
g	spectroscopic splitting factor, 2.0023
H	applied magnetic field; heat content
h	Planck's constant
Hz	hertz
I	spin quantum number; intensity; inductive effect
i.r.	infrared
J	coupling constant in n.m.r. spectra; joule
K	dissociation constant
kJ	kilojoule

LIST OF COMMON ABBREVIATIONS

k	Boltzmann constant; velocity constant
kcal	kilocalories
L	laevorotatory; laevo configuration
M	molecular weight; molar; mesomeric effect
Me	methyl
m	mass; mole; molecule; *meta-*
ml	millilitre
m.p.	melting point
Ms	mesyl (methanesulphonyl)
[M]	molecular rotation
N	Avogadro number; normal
nm	nanometre (10^{-9} metre)
n.m.r.	nuclear magnetic resonance
n	normal; refractive index; principal quantum number
o	*ortho-*
o.r.d.	optical rotatory dispersion
P	polarisation, probability; orbital state
Pr	propyl
Ph	phenyl
p	*para-*; orbital
p.m.r.	proton magnetic resonance
R	clockwise configuration
S	counterclockwise config.; entropy; net spin of incompleted electronic shells; orbital state
S_N1, S_N2	uni- and bi-molecular nucleophilic substitution mechanisms
S_Ni	internal nucleophilic substitution mechanisms
s	symmetrical; orbital; suprafacial
sec	secondary
soln.	solution
symm.	symmetrical
T	absolute temperature
Tosyl	p-toluenesulphonyl
Trityl	triphenylmethyl
t	time
temp.	temperature (in degrees centigrade)
tert.	tertiary
U	potential energy
u.v.	ultraviolet
v	velocity
Z	zusammen (together) configuration

LIST OF COMMON ABBREVIATIONS

α	optical rotation (in water unless otherwise stated)
$[\alpha]$	specific optical rotation
αA	atomic susceptibility
αE	electronic susceptibility
ε	dielectric constant; extinction coefficient
μ	microns (10^{-4} cm); dipole moment; magnetic moment
μB	Bohr magneton
μg	microgram ($10^{-6} g$)
λ	wavelength
ν	frequency; wave number
$\chi, \chi d, \chi \mu$	magnetic, diamagnetic and paramagnetic susceptibilities
$\sim$	about
(+)	dextrorotatory
(-)	laevorotatory
(±)	racemic
$\ominus$	negative charge
$\oplus$	positive charge

Chapter 24

THE CHEMISTRY OF PYRIDINE AND ITS DERIVATIVES

JAN BECHER

1. Introduction

This chapter covers the chemistry of pyridine and its derivatives from 1975 until the end of 1985. [Due to the wealth of information on pyridine which has been published over this ten years' period, only a limited selection of papers can be included in this review]. The arrangement of subjects follows that in chapter 24 of the second edition.

Review articles of various aspects of pyridine chemistry have appeared, the general ones will be mentioned here while the more specialized ones will be cited in the appropriate sections. Of special value is the compilation of reviews on pyridine chemistry ("Pyridine and Its Chemistry", Chapter IV, Part 14, ed. G. R. Newkome, in the series "The Chemistry of Heterocyclic Compounds", Vol. 5, ed. E. Klingsberg, Interscience, New York, 1984) which covers the period 1968-1982. Furthermore the same volume, chapter I, contains a comprehensive review on synthetic and natural sources of the pyridine ring (T. D. Bailey, G. L. Goe, and E. F. V. Scriven, *ibid.* , Chapter I).

Another important account of many aspects of pyridine chemistry is Vol. 2 of Comprehensive Heterocyclic Chemistry ("Comprehensive Heterocyclic Chemistry", Vol. 1-8, eds. A. R. Katritzky and C. W. Rees, Pergamon, 1985).

Acknowledgements. The author wishes to thank Inger Hansen for her patience and masterful typing of the manuscript, and also Edel Rasmussen who has elegantly drawn all of the illustrations.

2. Synthesis of pyridines

(a) From acyclic compounds

(i) Using ammonia or amines as sources of ring nitrogen

(1) Carbonyl compounds and ammonia*

In an alternative to the Hantzsch synthesis 1,4-dihydro-2,6-unsubstituted pyridines can be prepared in high yield via acetylenes (T. Chennat and V. Eisner, J. chem. Soc. Perkin I, 1975, 926).

$$2HC{\equiv}CCO_2Me \quad + \quad ArCHO \quad \xrightarrow{NH_4^{\oplus}} \quad \text{[structure]} \quad 52\text{–}79\%$$

The use of HMPT (hexamethylphosphoric triamide) in the preparation of pyridines in modest yields has been reported (R. S. Monson and A. Baraze, J. org. Chem., 1975, 40, 1672).

$$ArCOCHOHAr \quad \xrightarrow[\Delta]{HMPT} \quad \text{[structure]} \quad 10\text{–}19\%$$

By the Kröhnke method (P. Wild and F. Kröhnke, Ann., 1975, 849; for a review of this method see F. Kröhnke, Synthesis, 1976, 1) it is possible to prepare 3,5-bispyridinium salts.

$$2 \;\text{[structure]} \quad + \quad R^1CHO \quad \xrightarrow{NH_3} \quad \text{[structure]} \quad \xrightarrow{[O]} \quad \text{[structure]}$$

ca. 50%   ca. 90%

The use of palladium as catalyst for the preparation of 2-pyridones from 5-methyl- or 5-phenyl-2,4-pentadienamide has been described (A. Kasahara and T. Saito, Chem. and Ind., 1975, 745). Gas phase syntheses of pyridines on a platinum-silica catalyst and ammonia from diethyl ketone yield 2,6-diethyl-3- methylpyridine while methyl ethyl ketone gives a mixture of 2,3,6-trimethylpyridine and 2-methyl-6-ethyl-

---

*An access to the Russian literature on this subject can be found in (I. V. Lazdinsh and A. A. Avots, Chem. heterocyclic Compounds USSR, 1979, 823).

pyridine.

(2) Hantzsch type synthesis

Reviews on dihydropyridines have been published: (J. P. Kutney, Heterocycles, 1977, 7, 593; D. M. Stout and A. I. Meyers, Chem. Rev., 1982, 82, 223; J. Kuthan and A. Kurfürst, Ind. Eng. Chem. Prod. Res. Dev., 1982, 21, 191; F. Bossert, H. Meyer, and E. Wehinger, Angew. Chem. intern. Edn., 1981, 20, 762). The Hantzsch synthesis has received particular attention as some of the compounds formed are strongly pharmacologically active (F. Bossert and W. Vater, Naturwiss., 1972, 58, 578). Low yields are generally obtained with sterically hindered aldehydes, running the reaction under pressure is a useful modification (Y. Watanabe *et al.*, Synthesis, 1983, 761).

37–92%

A large number of dihydropyridines have been described in a series of papers (H. Meyer, F. Bossert, and H. Horstmann, Ann., 1978, 1476 and references cited therein). An example is given below.

18–62%

(3) Other condensations of dicarbonyl and α,β-unsaturated carbonyl compounds with reactive methylene compounds and ammonia

(A) Condensations of dicarbonyl compounds

The possible combinations of dicarbonyl compounds derivatives and active methylene compounds are many. A typical example is the reaction of acetylacetone and acetoacetamide (T. Kato and M. Noda, Chem. pharm. Bull., 1976, 24, 303).

For other examples see (T. Kato *et al.*, *ibid.*, 1980, <u>28</u>, 224).
   Condensation of 2-benzoylacetamidine with derivatives of ma-
lonodialdehyde and 1,3-ketoaldehydes yields 2-amino-3-pyridyl
phenyl ketones (M. Söllhuber-Kretzer and R. Troschütz, Arch.
Pharm., 1982, <u>315</u>, 783). Heating of simple enamino ketone
salts* gives high yields of 3-acylpyridines (S. Auricchio, R.
Bernardi, and A. Ricca, Tetrahedron Letters, 1976, 4831).

Nicotinamide derivatives can be prepared from ethyl 3-ethoxy-
methylene-2,4-dioxovalerate and β-aminocrotonamide (T. Kurihara and T. Uno, Heterocycles, 1977, <u>6</u>, 547). 1,4-Dihydro-4-
oxonicotinic acid derivatives are often biologically active
compounds. An alternative synthesis of this ring system is the
molecular sieve mediated cyclisation of β-keto esters and
β-aminocrotonates (O. Makabe, Y. Murai, and S. Fukatsu,
Heterocycles, 1979, <u>13</u>, 239).

Related preparations of 4-piperidones have been carried out by
condensation of acetone dicarboxylic esters, an aldehyde and
an amine (R. Caujolle, P. Castera, and A. Lattes, Bull. Soc.
Chim. France, 1983, 52).

---

*For preparation of some pyridines from enaminoketones, see J.
V. Greenhill, Chem. soc. Rev., 1977, <u>6</u>, 277; E. Stark *et al.*,
Chemiker Zeitung, 1977, <u>101</u>, 161. For gas-phase syntheses of
pyridines from acrolein, see H. Beschke and H. Friedrich,
ibid., 1977, <u>101</u>, 377.

(B) Condensations of α,β-unsaturated carbonyl compounds

This type of pyridine syntheses is very versatile and may be carried out with a large number of very varied starting materials, for example extension of the Kröhnke pyridine synthesis can give triarylpyridines (S. Malik *et al.*, J. chem. Eng. Data, 1983, 28, 430).

See also (R. S. Tewari and A. K. Dubey, *ibid.*, 1980, 25, 91) for the use of isoquinolinium ylides in the same type of synthesis.

2,4-Bis(dialkylamino)pyridines can be prepared from primary amines and trimethine cyanines (H. G. Viehe, G. J. de Voghel, and F. Smets, Chimia, 1976, 30, 189; M. H. Francotte, Z. Janousek, and H. G. Viehe, J. chem. Res. (S), 1977, 100).

Syntheses of 2,4,6-triarylpyridines usually give good yields. By the use of a furan ring instead of an aryl ring it is possible to eliminate the furan ring after ring closure by oxidation and decarboxylation (P. M. Carbateas and G. L. Williams, J. heterocyclic Chem., 1974, 11, 819; D. D. Weller, G. R. Luellen, and D. L. Weller, J. org. Chem., 1982, 47, 4803).

The use of sulphonium ylides in this type of syntheses has been described (R. S. Tewari and A. K. Awasthi, Synthesis, 1981, 314).

50–80%

The use of unsaturated esters and N,N-dialkylaminoacetals yields 4(1H)-pyridones (V. G. Granik *et al.*, Chem. heterocyclic Comp. USSR, 1976, 672. Ethyl 3,5,5-triethoxy-3-pentenoate and ammonia yield a 2(1H)-pyridone (P. D. Cook, R. T. Day, and R. K. Robins, J. heterocyclic Chem., 1977, $\underline{14}$, 1295).

93%

(4) Cyclisation of a C_5 chain with ammonia and amines

When the required starting material is readily available, cyclisation of C_5 compounds is an effective and selective method for preparation of pyridines. Cyclisation of an optically active dialdehyde and hydroxylamine yields an optically active pyridine via a Wittig reaction (G. Asara *et al.*, Synth. Comm., 1983, $\underline{13}$, 1129; C. Botteghi, G. Caccia, and S. Gladiali, Synth. Comm., 1976, $\underline{6}$, 549).

70–75%   40–60%

A simple synthesis from 1,6-heptadiene yields mixtures of Z- and E-2,6-dimethylpiperidines (J. Barluenga, C. Najera, and M. Yus, Synthesis, 1979, 896).

44–76%
(Z and E)

C_5-Enamino ketones prepared by acylation of enamines followed by reaction with amines yield 4(1H)-pyridones (R. F. Abdulla, T. L. Emmick, and H. M. Taylor, Synth. Comm., 1977, 7, 305).

66%

γ-Unsaturated ketoximes give pyridines when treated with a palladium catalyst (T. Hosokawa *et al.*, Tetrahedron Letters, 1976, 383). By the methodology previously used (B. Bak, G. O. Sørensen, and L. Mahler, Acta Chem. Scand., 1965, 19, 2001) ^{15}N has been incorporated in nicotinamide *via* the Zincke reaction* by the reaction of ^{15}NH$_3$ and a 1-(2,4-dinitro-phenyl)pyridinium halide (N. J. Oppenheimer, T. O. Matsunaga, and B. L. Kam, J. lab. Comp. Radiopharm., 1978, 15, 191).

(5) Other C_5 chain cyclisations

A number of piperidine syntheses *via* C_5-synthons have been reported. L-Piperidine-2-carboxylic acid can be prepared in one step from L-lysine by oxidation with disodium nitrosylpen-tacyanoferrate(II) (L. Kisfaludy and F. Korenczki, Synthesis, 1982, 163). Functionalized reduced pyridines can be prepared *via* Dieckmann cyclisation (J. Boujoch, I. Serret, and J. Bosch, Tetrahedron, 1984, 40, 2505).

Ring chain tautomerism of some 2-piperidinones prepared from glutarimides and phenyllithium has been described (Z. Czarnocki and J. T. Wrobel, Bull. Pol. Acad. Sci., 1984, 32, 335). The synthesis of 1,2-dihydropyridines, 2,3-dihy-dro-4(1H)-pyridinone, and 1,2,3,4-tetrahydropyridines from cyclisation of δ-acetylamino-α,β-ethylenic or acetylenic aldehydes is a useful method (J. P. Roduit and H. Wyler, Helv., 1985, 68, 403).

---

*A review of the Zincke aldehydes has been published (J. Becher, Synthesis, 1980, 589).

γ–Unsaturated amines can be cyclised with Lewis acids, for example $TiCl_3$. (J. L. Stein, L. Stella, and J. M. Surzur, Tetrahedron Letters, 1980, 21, 387) or other catalysts such as mercury salts (W. Carruthers, M. J. Williams, and M. T. Cox, Chem. Comm., 1984, 1235; R. Fuks, R. Merenyi, and H. G. Viehe, Bull. Soc. chim. Belg., 1976, 85, 147) in a related reaction alleneamines have been used as starting materials for the preparation of useful piperidines (S. Arseniyadis and J. Gore, Tetrahedron Letters, 1983, 24, 3997).

Mannich type cyclisations of aminoketals can give piperidines useful for the preparation of alkaloides (J. Bosch et al., J. heterocyclic Chem., 1983, 20, 595).

New methods for the preparation of pyridone derivatives have been reported, for example the conversion of acylanilides into 2–iminopyridines via Vielsmeier formylation (O. Meth-Cohn and K. T. Westwood, J. chem. Soc. Perkin I, 1983, 2089).

Pyrolysis of N–alkyl β–enaminoesters yields 3–formyl–4(1H)–pyridones (F. Arya, J. Boquant, and J. Chuche, Synthesis, 1983, 946).

An interesting C_5-method based on the versatile ketene dithio-acetals has been reported (K. T. Potts *et al.*, J. org. Chem., 1982, <u>47</u>, 3027 and references cited therein).

R, R^1 = aryl, thienyl etc.

2-Azabuta-1,3-dienes do not undergo a Diels-Alder reaction with acrylonitriles but react *via* a Michael addition. Reductive cyclisation of the adduct then gives a piperidine (H. Feichtinger, W. Payer, and B. Cornils, Ber., 1978, <u>111</u>, 1721).

69–95%

(ii) Using a cyano-group as the source of the ring nitrogen

(1) From hydrogen cyanide and other organic cyanides

2-Vinylpyridine may be obtained from acetylene and acryloni-trile with a cobalt catalyst (H. Bönnemann and M. Samson, German Patent 2840460, CA; 1980, <u>93</u>:95136). See also (H. Bönnemann, Angew. Chem. intern. Edn., 1978, <u>17</u>, 505). Annelated pyridines can be prepared by cobalt catalysis in related reactions in one step (A. Naiman, K. P. C. Vollhardt, Angew. Chem., 1977, <u>89</u>, 758; D. J. Brien, A. Naiman, and K. P. C. Vollhardt, Chem. Comm., 1982, 133), for a related reaction see also (F. A. Selimov, V. R. Khafizov, and U. M. Dzhemiler, Bull. Acad. Sci. USSR, 1983, 1709).

n = 3,4,5

This type of reaction can also be used for the preparation of optically active pyridines from an optically active nitrile and an acetylene (T. Tatone *et al.*, J. org. Chem., 1975, <u>40</u>, 2987). Cobalt-catalysts also give 2-alkylthiopyridines from acetylenes and alkyl thiocyanates (H. Bönnemann and G. S.

Natarajan, Erdöl und Kohle, Erdgas, Petrochemie, 1980, <u>33</u>, 328).

$$RC{\equiv}CR + R^1SCN \xrightarrow{\text{Co/cat.}}$$

75–89%, R = C_6H_5, R^1 = alkyl

A Diels–Alder reaction of "push–pull" activated isoprenes gives pyridines (M. Gillard *et al.*, J. Amer. chem. Soc., 1979, <u>101</u>, 5837).

$$+ \ RCN \longrightarrow$$

60–62%

Various anions can react with organic nitriles to give pyridines. Thus, lithiated imines give pyridines (K. Takabe *et al.*, Tetrahedron Letters, 1975, 4375).

$$\xrightarrow{\text{LDA}} \quad \left[\quad \right] \xrightarrow{R^1CN}$$

22–63%

R = H. alkyl

The acetylacetonitrile dianion with a number of nitriles, such as benzonitrile, yields 4(1H)-pyridones (F. J. Vinick, Y. Pan, and H. W. Gschwend, Tetrahedron Letters, 1978, 4221).

$$\xrightarrow{\text{LDA}} \xrightarrow{\text{ArCN}}$$

48–67%

In an interesting reaction the trimethylene methane dianion and benzonitrile give a pyridine (R. B. Bates *et al.*, J. org. Chem., 1980, <u>45</u>, 168).

$$\left[\quad \right]^{2\ominus} \xrightarrow{\text{PhCN}}$$

85%

(2) From Malononitrile

Malononitrile and its derivatives are versatile starting materials for a large number of pyridine syntheses. Alkylidene cyanoacetates react with formamides to give unsaturated aldehyde derivatives which can be cyclised to nicotinic acid derivatives (J. J. Baldwin, A. W. Raab, and G. S. Pouticello, J. org. Chem., 1978, 43, 2529).

2-Methylenepyridines can be prepared in a related synthesis from the malononitrile dimer and benzylidene malononitriles (L. Fuentes and J. J. Vaquero and J. L. Soto, Synthesis, 1982, 320). [2+4]- Cycloaddition of conjugated dienes and p-tosylisonitrosomalononitrile gives pyridines (J. P. Fleury, M. Desbois, and J. See, Bull. Soc. chim. France, 1978, II-147).

3,5-Dicyanopyridines can be prepared from N,N-dimethyl acetamide diethyl acetal and malononitrile (S. I. Kaimanakova *et al.*, J. org. Chem. USSR, 1983, 19, 988). A number of 3-cyanopyridine syntheses using aldehydes, ketones, and malononitriles have been described (S. Kambe *et al.*, Synthesis, 1980, 366).

This method has been varied by the use of chalcones and malononitrile derivatives (S. M. A. D. Zayed and A. Attia, J. heterocyclic Chem., 1983, 20, 129, and references cited therein). In the same type of synthesis 2-cyanocinnamates and malononitrile yield 2(1H)-pyridones (H. H. Otto, O. Rinus, and H. Smelz, Synthesis, 1978, 681). If this reaction is carried out

with sodium ethanethiolate, 6-ethylthiopyridines are obtained
(L. Fuentes, J. J. Vaquero, and J. L. Soto, J. heterocyclic
Chem., 1982, 19, 1109 and references cited therein).

In a related synthesis β-oxo thioamides and malononitrile with
piperidine as catalyst yield 2(1H)-pyridinethione derivatives
(K. B. Szwed, Monatsh., 1982, 113, 583). Related 2(1H)-pyri-
dinethiones can be obtained from ketene-S,N-acetals (H.
Takahata, T. Nakajima, and T. Yamazaki, Chem. pharm. Bull.,
1984, 32, 1658).

(3) From glutaronitriles (1,3-dicyanopropanes)

The dimer of acrylonitrile can be used for preparation of
halogenomethylpyridines (H. Fritz, C. D. Weis, and T. Winkler,
Helv., 1976, 59, 179).

*(4) Pyridines from other types of nitriles, cyanoaceta-
mides etc.*

A large number of 2(1H)-pyridone derivatives have been prepared by cyclisation reactions of cyanoacetamides, cyanoacetates, and related compounds involving Michael type additions
and basic condensations, such as the Knoevenagel reaction (R.
Baker et al., J. chem. Soc. Perkin I, 1979, 677); The reactions are dependent upon the base concentration and the yields
are in the range 61-79%. A representative example is given
below.

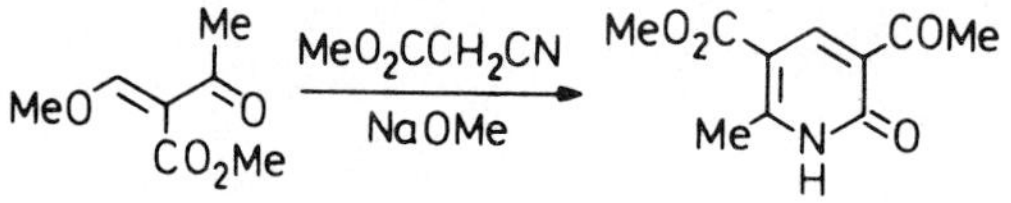

Other useful examples can be found in the following references: M. A. Cabrerizo and J. L. Soto, Ann. Quim., 1976, 72,
926; C. Seoane, J. L. Soto, and M. P. Zamorano, Heterocycles,
1980, 14, 639; J. L. Soto, C. Seoane, and A. M. Mansilla, Org.
Chem. Prep. Procedures, 1981, 13, 331; S. Kambe *et al.*, Synthesis, 1977, 841; 1981, 211; W. Jünemann, H. J. Opgenorth,
and H. Schevermann, Angew. Chem., 1980, 92, 390; V. S.
Hawaldar and S. V. Sunthankar, Ind. J. Chem., 1980, 19B, 151;
J. L. Soto *et al*, Synthesis, 1981, 529; M. J. Rubio, C.
Seoane, and J. L. Soto, Heterocycles, 1983, 20, 783; G. E. H.
Elgemeie *et al.*, Heterocycles, 1984, 22, 2829. Ethyl pro-
piolate and cyanacetamide give 2,6-dihydroxypyridines (W.
Jünemann, H. J. Opgenorth, and H. Scheuermannn, Angew. Chem.,
1980, 92, 390).

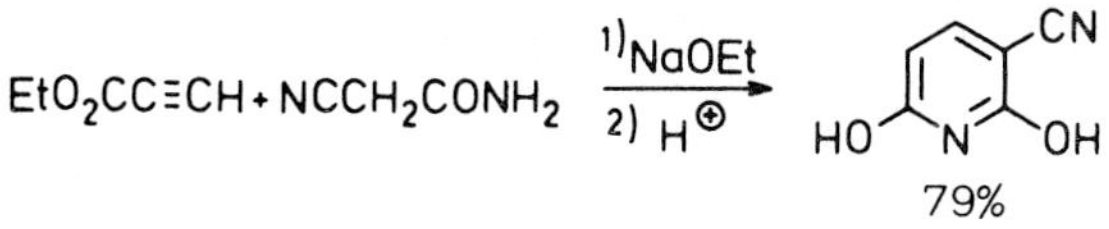

An important variation is the use of ketene-S,S-acetals
easily prepared from acetophenones (R. R. Rastogi, A. Kumar,
H. Ila, and H. Junjappa, J. chem. Soc. Perkin I, 1978, 554 and
references cited therein).

A number of aminopyridine derivatives can be obtained from
malononitrile and malonoamide derivatives (G. Koitz, W.
Fabian, H. W. Schmidt, and H. Junek, Monatsh., 1981, 112, 973
and references cited therein).

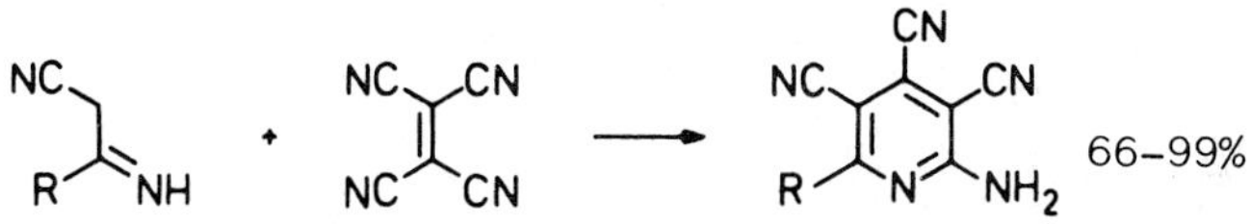

R = alkyl, aryl

The following reaction (S. K. Robev, Tetrahedron Letters,
1980, 21, 2097) is a preparation of the virtually unknown
2,5-dihydropyridine system *via* a Thorpe reaction.

Z and *E*

Enamines easily prepared from β-aminocrotonic esters and cyanoacetic acid derivatives give 2-amino-4(1*H*)-pyridones (T. Kappe, H. P. Stelzel, and E. Ziegler, Monatsh., 1983, 114, 953) while 3-alkylthiopyridines may be prepared from various 1-cyano-1-akylthio ethylenes (F. Pochat, Tetrahedron Letters, 1983, 24, 5073). 2-Amino or 2-chloro-3-cyanopyridines can be prepared from enamines and malononitrile derivatives (H. Kurihara and H. Mishima, J. heterocyclic Chem., 1977, 14, 1077).

The same reaction can also be carried out intramolecularly (G. Ege, H. O. Frey, and E. Schuck, Synthesis, 1979, 376).

Dimerisation of *N*-alkylcyanoacetamides with ethoxide yields 2(1*H*)-pyridones (E. Schmitz and S. Schramm, J. prakt. Chem., 1982, 324, 82).

(iii) Cyclisations not involving C–N bond formation

(1) Cycloaddition reactions of isocyanates, isothiocyanates, and other heterocumulenes

Enaminoketones and benzoyl isothiocyanate give moderate yields of 2(1*H*)-pyridinethiones (O. Tsuge and A. Inaba, Heterocycles, 1975, 3, 1081).

$$\text{Me, COPh} \quad + \quad \text{PhCONCS} \quad \longrightarrow \quad \text{(NR}_2\text{, PhCO, S, N, Ph, COPh)} \quad + \quad \text{(NR}_2\text{, PhCO, COPh, S, N, H, NHCOPh)}$$

42%   27%

A general method for the preparation of 3-acyl or 3-for-
myl-2(1H)-pyridinethiones (as well as pyridineselones and pyridones) from the glutaconaldehyde anion or related 1,5-pen-
tenedione anions and organic isothiocyanates (isoselenocya-
nates and isocyanates) has been described (J. Becher *et al.*,
Acta Chem. Scand., 1977, B31, 843; F. M. Asaad and J. Becher,
Synthesis, 1983, 1025; J. Becher, F. M. Asaad, and I.
Winckelmann, Ann., 1985, 620 and references cited therein).
The yields are usually in the range 50-99%; the starting enol
anions can be obtained in fair yields from ring opening of
pyridinium or pyrylium salts.

$$R^1, R^3, R^5 = H,\ \text{alkyl, aryl;} \quad X = S,\ Se\ \text{or}\ O$$

Cycloaddition of vinyl-heterocumulenes and ynamines gives
4(1H)-pyridones or thiones together with derivatives of these
products (A. Dondoni, L. Kniezo, and A. Medici, J. org. Chem.,
1982, 47, 3994).

20%   3.7%

Nickel or cobalt catalysed reactions of acetylenes and isocyanates yield 2(1H)-pyridones in high yields (H. Hoberg and G.

Burkhart, Synthesis, 1979, 525; H. Hoberg and B. W. Oster, *ibid.*, 1982, 324).

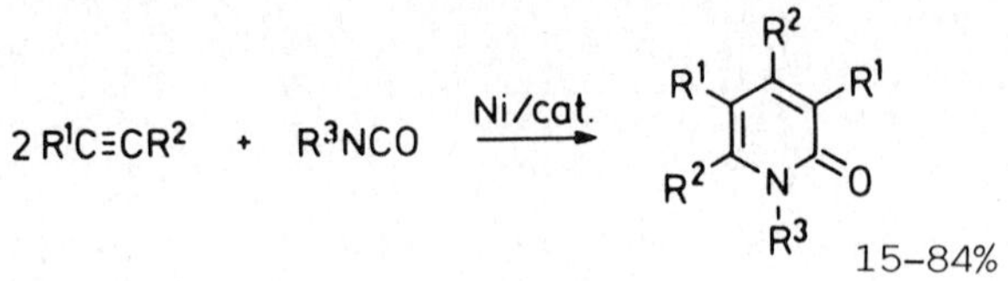

$$2\,R^1C{\equiv}CR^2 \;+\; R^3NCO \xrightarrow{\text{Ni/cat.}}$$

15–84%

In a similar reaction carbodiimides give 2(1H)-iminopyridines (P. Hony and H. Yamazaki, Tetrahedron Letters, 1977, 1333).

(2) Cycloaddition reactions of imines and acetylenes

Various types of imines and acetylenes can give pyridines (J. Barluenga, M. Tomas, S. Fustero, and V. Gotor, Synthesis, 1979, 345 and references cited therein).

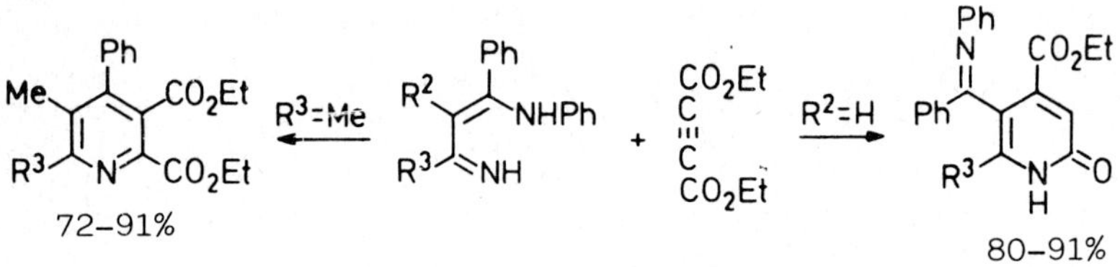

72–91%

80–91%

2-Pyridones can be prepared by thermal rearrangements of pyrrolidine pseudoureas (L. E. Overman *et al.*, J. Amer. chem. Soc., 1980, 102, 747; L. E. Overman and J. P. Roos, J. org. Chem., 1981, 46, 811).

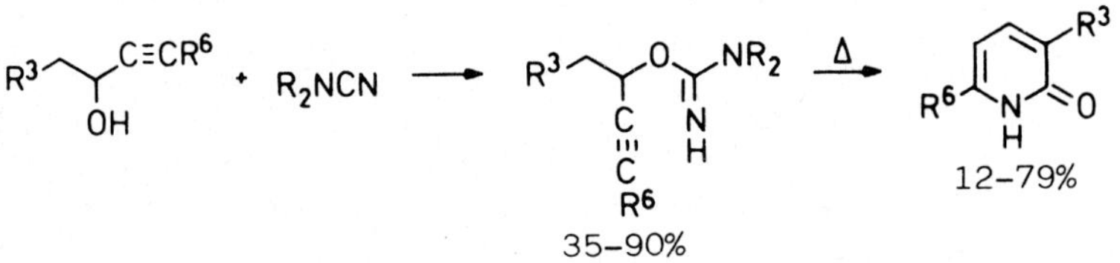

35–90%

12–79%

Imidoyloxosulphonium ylides react with electrophilic acetylenes yielding 1,2-dihydropyridines (R. Faragher, T. L. Gilchrist, and I. W. Southon, Tetrahedron Letters, 1979, 4113).

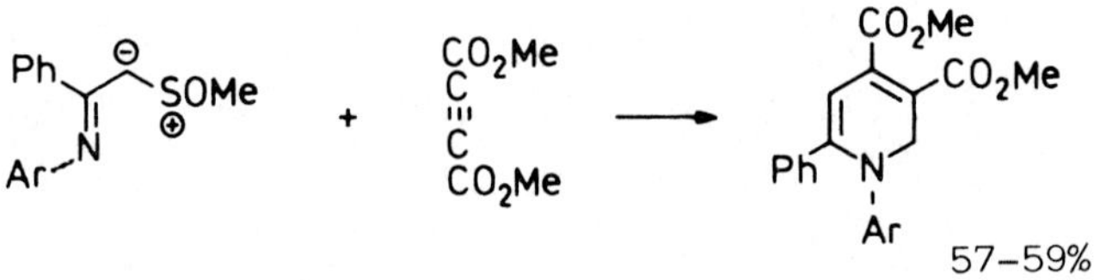

57–59%

Azabutadienes are versatile dienes undergoing cycloadditions with acetylenes to yield pyridines (F. Sainte *et al.*, J. Amer. chem. Soc., 1982, 104, 1428; R. Gompper and U. Heinemann, Angew. Chem., 1980, 92, 208; 1981, 93, 287).

(3) Other cycloaddition reactions

[4+2]-Cycloadditions using various dienes and ene systems can be used for construction of reduced pyridines. A general and simple synthesis of tetrahydropyridines uses cyclocondensation of dienes with simple iminium salts generated *in situ* under Mannich conditions (S. D. Larsen and P. A. Grieco, J. Amer. chem. Soc., 1985, 107, 1768).

Other examples of cyclocondensation of dienes and imines giving reduced pyridines can be found in the following references. M. E. Jung *et al.*, Tetrahedron Letters, 1981, 22, 4607; A. N. Mirskova *et al.*, J. org. Chem. USSR, 1983, 19, 1547; K. Krishan, A. Singh, and S. Kumar, Synthetic Comm., 1984, 14, 219; D. Prajapati *et al.*, Heterocycles, 1984, 22, 287. Photochemical cycloaddition of vinylogous formamides can give 1,4-dihydronicotinic acids (L. F. Tietze and K. Brüggemann, Angew. Chem., 1979, 91, 575).

2(1*H*)-Pyridones can be prepared from 1-azabutadienes and ester enolates (M. Komatsu *et al.*, Tetrahedron Letters, 1981, 3769).

27–78%

Fully aromatic pyridines can be prepared by cyclocondensation of enamines with N-methylene-*tert*-butylamine in 67–87% yields (M. Komatsu *et al.*, Angew. Chem., 1982, 94, 214). Vinamidinium salts* and β-aminocrotonates can be cyclised to pyridines (J. C. Jutz, H. G. Löbering, and K. H. Trinkl, Synthesis, 1977, 326).

R = $-CO_2R''$, CN; R^1 = H, alkyl, methoxy, phenyl   52–95%

Annelated pyridines can also be obtained by the method described above as well as by cycloaddition reactions of N-(1-cyanoethyl)formamide and dienophiles, such as N-phenyl-maleimide resulting in the formation of pyridoxine derivatives (S. Shimada and M. Oki, Chem. pharm. Bull., 1984, 32, 38). Thermolyses of cycloalkanone oxime O-allyl ethers in air give the corresponding cycloalkenopyridines in 30–65% yields (J. Koyama *et al.*, *ibid.*, 1983, 31, 2601 and references cited therein). Interesting bis-annelated pyridines can be prepared from homoadamantanone N,N-dimethylhydrazonium iodide (R. E. Parton *et al.*, Tetrahedron Letters, 1982, 4447).

(b) From other heterocyclic compounds

(i) From 5-membered heterocyclic ring systems

(1) Furans

A useful high temperature modification of the Clauson-Kaas reaction for 3-hydroxypyridines has been reported, 2-amino-3-hydroxypyridines can be prepared from substituted 2-furoic

---

*For a review of other pyridine syntheses *via* electrocyclic ring closure with elimination, see J. C. Jutz, Topics in Current Chem., 1978, 73, 125.

acids (H. Greuter and D. Bellus, J. heterocyclic Chem., 1977, 14, 203).

R^1 = H or alkyl, R^2 = $CONH_2$, $COOEt$, $COOH$ etc.

(2) Pyrroles

An interesting ring expansion of pyrrolin-2,3-diones with diazoalkanes has been reported (B. Eistert, G. W. Müller, and T. J. Arackal, Ann., 1976, 1023).

(3) Oxazoles, thiazoles, and selenazoles

A modification of the well known Diels-Alder approach to pyridoxal derivatives from silylated 2-oxazolin-5-ones has been described (H. Takagaki, N. Yasuda et al., Chem. Letters, 1979, 183).

Oxazoles have been suggested as intermediates in the preparation of 3-aminopyridines in the cycloadditions of N-(cyano-phenylmethyl)acylamides or 2-acylamino-2-cyanoacetates with olefins (S. Shimada and H. Maeda, Chem. pharm. Bull., 1983, 31, 3460).

Mesoionic selenium azoles can also undergo cycloadditions to give pyridines (M. P. Cava and L. E. Saris, Chem. Comm., 1975, 617).

The corresponding sulphur azoles yield thiophenes. Intramolecular Diels–Alder reactions have been carried out with oxazoles yielding pyridines (S. Götze, B. Kübel, and W. Steglich, Ber., 1976, 109, 2331).

Related intramolecular cycloadditions *via* an azabenzvalene intermediate take place with 3-cyclopropenyl oxazolinones (A. Padwa *et al.*, J. Amer. chem. Soc., 1982, 104, 286).

(4) Isoxazoles

A molybdenum catalysed cycloaddition of isoxazoles and acetylene dicarboxylate gives a pyridine analogue of [6]pyridinophane (N. Nitta and T. Kobayashi, Tetrahedron Letters, 1984, 959).

(5) Pyridines from other 5-membered heterocyclic rings

Triazoles give rise to azabutadienes which, by reaction with acetylenes, give pyridines (see also *(a) (iii) (2)*) (Y. Nomura *et al.*, Chem. Letters, 1979, 187).

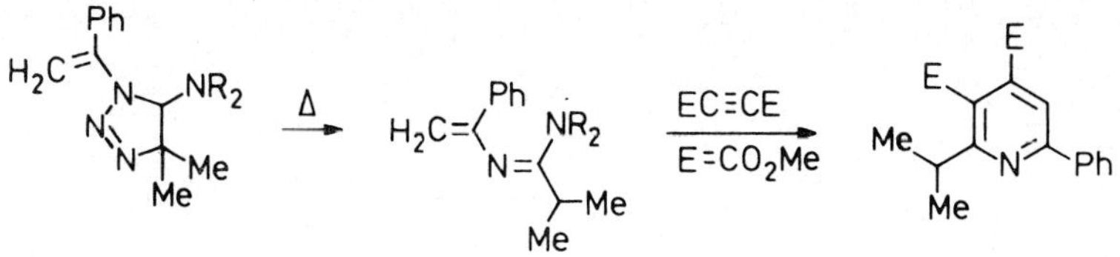

Triazolo[1,5-a]pyridine can be used for preparation of 2-for-mylpyridines (G. Jones and D. R. Sliskovic, Tetrahedron Letters, 1980, 21, 4529).

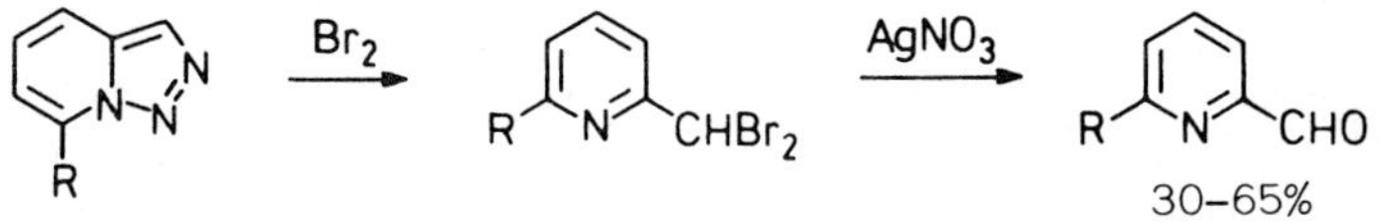

30–65%

6-Substituted pyridines can be obtained by alkylation of the starting triazolo[1,5-a]pyridine via lithiation. A pyrazoline ring may also give 2,4,6-triphenylpyridine (M. Lempert-Sreter and K. Lempert, Tetrahedron, 1975, 31, 1677).

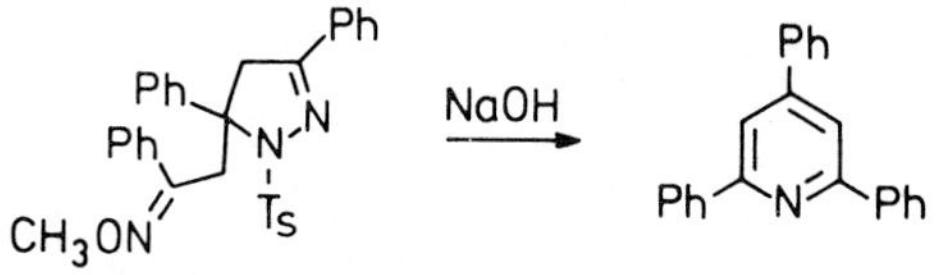

The potential synthetic utility of 1,2-fused pyridines is illustrated in the following examples, where the new azetepy-ridinium system is obtained (P. J. Pointer and J. B. Wilford, J. chem. Soc. Perkin II, 1983, 403).

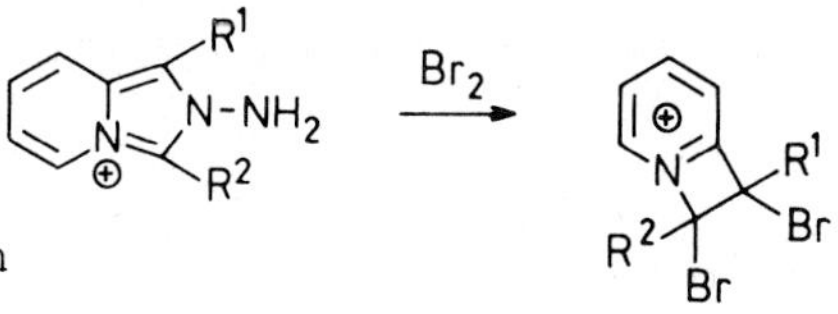

R^1, R^2 = Ph

3-Nitroso imidazo[1,2-a]pyridines give pyridines by phosphite reduction (D. J. Birch et al., J. org. Chem., 1982, 47, 3547).

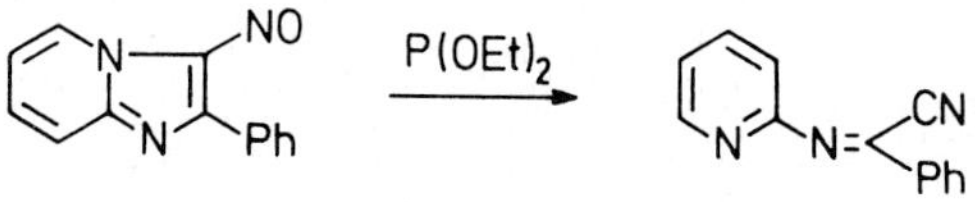

Stable pyridinium dinitromethylides can be obtained in high

yields from imidazo[1,2-*a*]pyridines (E. Andreasson *et al.*, Chem. Comm., 1983, 816).

A 2-piperidone synthesis starting from an azido substituted lactone has been reported (R. K. Olsen, K. L. Bhat, and R. B. Wardle, J. org. Chem., 1985, 50, 896).

(ii) From 6-membered heterocyclic ring systems

(1) Pyrylium salts

The conversion of pyrylium salts into pyridines with ammonia and primary amines has been reviewed (A. T. Balaban, A. Dinculescu, G. N. Dorofeenko, G. W. Fischer, A. V. Koblik, V. V. Mezheritski, and W. Schroth, Adv. heterocyclic Chem. Suppl. 2, "Pyrylium Salts", Academic Press, New York, 1982). The use of pyrylium salts for conversion of amines into other functionalities *via* pyridinium compounds has been reported in a number of papers, for a review see A. R. Katritzky, Tetrahedron, 1980, 36, 679 and references cited therein.

1-Substituted-2,4,6-triarylpyridinium salts have been used extensively, but the corresponding tetrahydroacridinium salts are generally better leaving groups and are therefore more reactive, (A. R. Katritzky, J. chem. Soc. Perkin I, 1981, 1492).

Another interesting example is the following rearrangement (A. R. Katritzky and R. Awartani, J. chem. Soc. Perkin I, 1983, 2623).

Another modification is the use of water soluble pyrylium and pyridinium salts (A. R. Katritzky *et al.*, J. chem. Soc. Perkin II, 1984, 857). The sterically hindered base, 2,6-di- *tert* -butylpyridine, is prepared from the corresponding pyrylium salt, easily obtained from pivaloyl chloride and isobutene (A. G. Anderson and P. J. Stang, J. org. Chem., 1976, 41, 3034).

Other preparations of this type have been reported (R. Neidlein and P. Witerzens, Monatsh., 1975, 106, 643; J. P. Le Roux, J. C. Cherton, and P. L. Desbene, Compt. Rend., 1975, 280C, 37; A. T. Balaban, J. labelled Comp. and Radiopharm., 1981, 18, 1621). The preparation of 2,4,6-trimethylpyridinium ferrocenyl ylide has been reported (R. A. Abramovitch and W. D. Holcomb, J. org. Chem., 1976, 41, 41).

(2) Pyrones, thiopyrones, pyranes, and thiopyranes

The conversion of pyrones and related compounds to pyridines is a useful method for the preparation of 2- and 4(1 *H*)-pyridones.

For example pyridinium compounds can be prepared by the reaction of pyrones with *N* -aminopyridinium salts (A. R. Katritzky and M. P. Sammes, Chem. Comm., 1975, 247). β -Diketones and 2,4,6-trichlorophenyl cyanoacetate yield 2-pyrones and hence 2-pyridones with amines (E. Ziegler, F. Raninger, and A. H. Müller, Ann., 1976, 250). A large scale preparation of an important intermediate for methotrexate synthesis can be prepared from chelidonic acid (E. C. Taylor and J. S. Skotnichi, Synthesis, 1981, 606).

96%

The reaction of chelidonic acid and amines can give rise to a variety of products, such as: salts, N-arylchelidamic acids, N-aryl-4-pyridone-2-carboxylic acids, N-aryl-4-pyridones or chelidamic acid depending upon reaction conditions and the amine (A. R. Katritzky, R. Murugan, and K. Sakizadeh, J. heterocyclic Chem., 1984, 21, 1465).

In an alternative to the Hantzsch synthesis 1,4-dihydropyridines can be prepared from reduced pyrans (J. S. Foss et al., Tetrahedron Letters, 1978, 1407).

42%

N-Aminopyridines can be prepared from 2-pyrones and hydrazines (A. Roedig, J. Hilberth, and H. A. Renk, Ann., 1975, 2251).

Pyran-2-thiones can be useful starting materials for $2(1H)$-pyridinethiones (P. M. Fresneda et al., Synthesis, 1981, 711), while a Dimroth rearrangement in base can give $2(1H)$-pyridinethiones from a 2-imino dihydrothiopyran (K. Schweiger, Monatsh., 1983, 114, 581).

4-Methylenepyridines can be prepared from the corresponding pyrones (J. Ploquin et al., Eur. J. med. Chem., 1982, 17, 149; F. Eiden, M. Beuttenmüller, and H. Schaumburg, Arch. Pharm., 1975, 308, 489).

The reaction of 4-alkylidene pyrans with hydroxylamine may lead to pyridine *N*-oxides; however, depending upon reaction conditions isoxazoline derivatives can also be formed (P. Crabbé *et al.*, J. chem. Soc. Perkin I, 1975, 1342). For example *N*-substituted tetrahydropyridines can be prepared *via* a hetero-Cope rearrangement (K. B. Lipkowitz *et al.*, Tetrahedron Letters, 1979, 2241).

(3) Oxazines

Pyridines can be prepared from 1,2-oxazines (R. Faragher and T. L. Gilchrist, Chem. Comm., 1977, 252; T. L. Gilchrist, G. M. Iskander, and A. K. Yagoub, Chem. Comm., 1981, 696).

80%

Also pyridine *N*-oxides may be prepared.

1,3-Oxazin-4-ones are excellent starting materials for the preparation of pyridines (T. Kato, Y. Yamamoto, and M. Koudo, Chem. pharm. Bull., 1975, 23, 1873; T. Kato and M. Koudo, *ibid.*, 1976, 24, 356; Y. Yamamoto and Y. Morita, *ibid.*, 1985, 33, 975), for example by reaction with active methylene compounds and base.

Cycloaddition of a 1,3-oxazin-6-one with an acetylene yields a pyridine (S. Götze and W. Steglich, Ber., 1976, 109, 2327).

Aminomethylation of α-methylstyrene yields 1,2,3,6-tetrahydro-4-phenylpyridines by a hydrochloric acid-catalyzed rearrangement of an intermediate oxazine derivative (P. Sohar, J. Lazar, and G. Bernath, Ber., 1985, 118, 1985).

(4 and 5) Pyridazines and triazines

Dienophiles react with 1,2,4-triazones to give pyridines in short synthetic routes (M. G. Barlow, R. N. Hazeldine, and D. L. Simphin, Chem. Comm., 1979, 658; D. L. Boger and J. S. Panek, J. org. Chem., 1981, 46, 2179; D. L. Boger, J. S. Panek, and M. M. Meier, ibid., 1982, 47, 895).

Enamines generated in situ from ketones react under mild conditions with 1,2,4-triazines to give alkylpyridines in high yields (D. L. Boger and J. S. Panek, J. org. Chem., 1981, 46, 2179; D. L. Boger, J. S. Panek, and M. M. Meier, ibid., 1982, 47, 895).

1,3,5-Triazines react with ethyl acetoacetate anions to give ethyl 5-substituted-4-oxo-1,4-dihydro-3-pyridinecarboxylates in a nucleophilic ring opening and ring closure reaction (M. Balogh *et al.*, J. heterocyclic Chem., 1980, _17_, 359).

(6) Pyrimidines

Ring degenerate transformations of azines have been reviewed (H. C. Van der Plas, Tetrahedron, 1985, _41_, 237). Thus, a number of pyrimidine derivatives can give pyridines, for example nitriles of the type RCH_2CN (R = Ph, CN, COOR etc.) are effective reagents for conversion of 5-nitropyrimidine into 2-amino-5-nitro-3-substituted pyridines (V. N. Charushin and H. C. van der Plas, Rec. Trav. Chim., 1983, _102_, 372).

An interesting variation of this type of ANROC reaction is the transformation of 2(1H)-pyrimidones or thiones into pyridines with anions of active methylene compounds (A. Katoh, Y. Omote, and C. Kashima, Heterocycles, 1984, _22_, 763; Chem. pharm. Bull., 1984, _32_, 2942).

(iii) From other ring systems

In a thermal transformation allyl and other substituted 2H-azirines give pyridines (A. Padwa and P. H. J. Carlsen,

Tetrahedron Letters, 1978, 433; J. org. Chem., 1978, 43, 3757
and references cited therein).

A low yield of 2,3,5,6-tetraphenylpyridine from 3-phenyl-2-
styryl-2*H*-azirine with iron nonacarbonyl has been reported
(F. Bellamy, Chem. Comm., 1978, 998).
   Oxaziridines and enamines can give pyridines (M. Komatsu
et al., Tetrahedron Letters, 1976, 4589).

Diketene and primary amines yield acetoacetamides which in
turn react with another mol of diketene to give 2(1*H*)-pyri-
dones (P. Dimroth and V. Radtke, Ann., 1979, 769).

28–69%

In a ring opening reaction diazetidine carboxylic esters give
an *N*-aminopyridine (D. Mackay and L. L. Wong, Canad. J.
Chem., 1975, 53, 1973).

(c) From carbocyclic compounds

By a fairly general reaction cyclobutadienes, easily prepared
from enamines, can give derivatives of 2(1*H*)-pyridinethiones
(R. Gompper, S. Mensch, and G. Seybold, Angew. Chem., 1975,
87, 711).

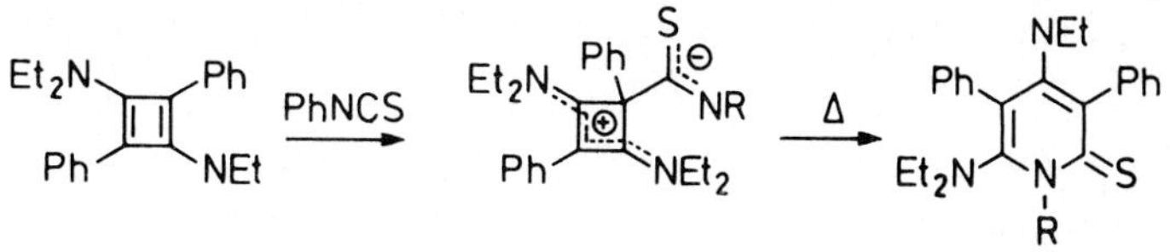

In a related reaction cyclobutadiene σ-complexes react with nitriles (P. B. J. Driessen *et al.*, Tetrahedron Letters, 1976, 2263; H. Hogeveen, R. F. Kingma, and D. M. Kok, J. org. Chem., 1982, **47**, 989).

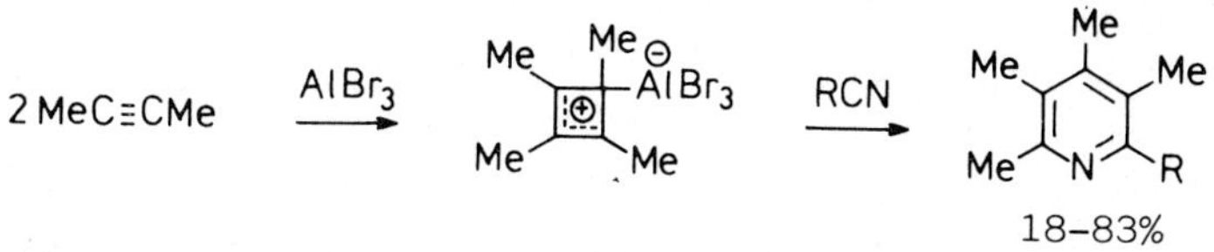

18–83%

Cyclopentane azides can be useful starting materials for alkaloids derived from piperidines (A. Astier and M. M. Plat, Tetrahedron Letters, 1978, 2051).

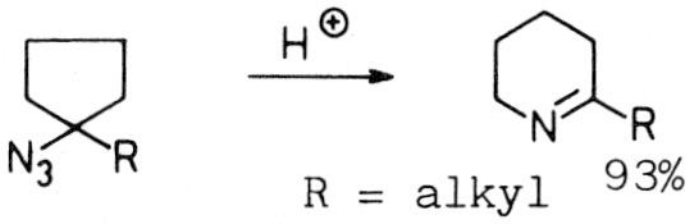

R = alkyl   93%

Cyclopentadienones react with azides to give pyridines in a cycloaddition reaction (A. Hassner, D. J. Anderson, and R. H. Reuss, Tetrahedron Letters, 1977, 2463).

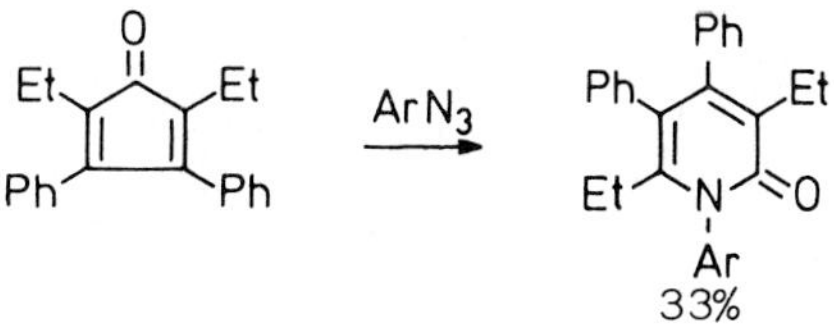

33%

3. Reaction of pyridines

(a) Pyridines as tertiary amines

pKa Values for a number of sterically hindered pyridines have been determined in methanol at 25°C (W. J. le Noble and T. Asano, J. org. Chem., 1975, **40**, 1179). The extraordinary combination of properties of 2,6-di-*t*-butylpyridine has been reviewed and compared to other pyridines (B. Kanner, Heterocycles, 1982, 18, 411). The following table gives a compilation of pKa values from the references cited above, see also the 2nd edition, page 158.

Pyridine	pKa
pyridine	4.38^a
2-methyl-	5.05^a
2-ethyl-	4.93^a
2-isopropyl-	4.82^a
2-t-butyl-	4.68^a
2,6-dimethyl-	5.77^a; 6.86^b
2,6-diethyl-	6.9^b
2,6-diisopropyl-	5.34^a; 6.6^b
2-methyl-6-t-butyl-	5.52^a
2-ethyl-6-t-butyl-	5.36^a
2-isopropyl-6-t-butyl-	5.13^a
2,6-di-t-butyl-	3.58^a; 4.2^b

[a] 50% EtOH-H_2O at 27° $\pm$ 2°C. [b] MeOH at 25°C.

Heats of reaction for a series of pyridines with alkyl fluorosulphonate have been reported, charge delocalisation in the pyridine ring was found to be an important factor (E. M. Arnett and C. Petro, J. Amer. chem. Soc., 1976, 98, 1468). For an account on the steric requirement of the nitrogen lone pair in pentaarylpyridines, see D. Gust and M. W. Fagan, J. org. Chem., 1980, 45, 2511. For an account on pyridine N_{1s} ionization energies, see B. S. Brown and A. Tse, Canad. J. Chem., 1980, 58, 694.

(i) Pyridinium salts

Pyridinium salts can be prepared by a number of methods as described in this section of the 2nd edition, for example by N-alkylation, N-acylation etc. A number of ring syntheses result in the formation of pyridinium salts. The quantitative aspects of quarternization of pyridines have been reviewed (J. A. Zoltewicz and L. W. Deady, Adv. heterocyclic Chem., 1978, 22, 71). One of the more important methods for the formation of pyridinium compounds is their preparation from pyrylium salts and amines. This method has been extensively used for the conversion of primary amines into a number of other functionalities as the pyridine moiety is an excellent leaving group, for a review see (A. R. Katritzky, Tetrahedron, 1980,

36, 679). A typical example is the conversion of primary amines into olefins (A. R. Katritzky and A. M. E. Mowafy, Chem. Comm., 1981, 96) or the preparation of bromides from primary amines (A. R. Katritzky et al., J. chem. Soc. Perkin I, 1980, 1890). The general scheme for this type of reaction is given below.

Normally N-methylation of pyridines gives poor yields with diazomethane in the absence of tetrafluoroboric acid, but 2-pyridylphosphonic acid can be quantitatively N-methylated in a direct reaction with diazomethane (J. S. Loran, R. A. Naylor, and A. Williams, J. chem. Soc. Perkin II, 1976, 1444). An interesting poly-N-pyridinium salt has been prepared from tetrachlorocyclopropene and 4-(dimethylamino)pyridine (K. C. Waterman and A. Streitwieser, J. Amer. chem. Soc., 1984, 106, 3874).

In a series of papers Mukaiyama et al. have reported the use of pyridinium compounds as reagents and catalysts in a number of synthetically useful reactions, for a review see (T. Mukaiyama, Angew. Chem. intern. Edn., 1979, 18, 707), an interesting example is the preparation of macrocyclic lactones in high yields (73–99%) (T. Mukaiyama, K. Narasaka, and K. Kikuchi, Chem. Letters, 1977, 441).

Nitration via N-nitropyridinium salts has been extended to transfer nitration of alcohols (G. A. Olah *et al.*, Synthesis, 1978, 452).

$$R\text{-}OH \ + \ \ \longrightarrow \ RONO_2 \ + $$

38–100%

Aromatic compounds may also be nitrated in fair yields by this methodology using 2,4,6-trimethylpyridine (G. A. Olah *et al.*, J. Amer. chem. Soc., 1980, <u>102</u>, 3507). Sulphonation of sensitive hydroxy compounds can be carried out *via* a pyridinium compound (E. Anders and A. Stankowiak, Synthesis, 1984, 1039).

63–91%

N-Alkyl salts of 4-dialkylaminopyridines are effective phase transfer catalysts which are thermally stable (D. J. Brunelle and D. A. Singleton, Tetrahedron Letters, 1984, 3382). Pyridine, acyl chlorides, and aldehydes react at room temperature to give pyridinium compounds which are useful for the preparation of esters (E. Anders and T. Grassner, Angew. Chem., 1982, <u>94</u>, 293) and (E. Anders, W. Will, and T. Grassner, Ber., 1983, <u>116</u>, 1506).

The preparation of pyridinium dichromate reagent under aqueous conditions has been reported to be hazardous resulting in an explosion (J. Salmon, Chem. Industry, 1982, 616). Alkylation of 4-(4-nitrobenzyl)pyridine with carcinogenic alkylating agents yields two types of products depending upon the reagent used (B. M. Goldschmidt, B. L. van Duuren, and R. C. Goldstein, J. heterocyclic Chem., 1976, <u>13</u>, 517).

(ii) Pyridine N-oxides*

(1) Preparation

Selective ring nitrogen oxidation of pyridines with an unsubstituted 2-amino moiety can be carried out with 3-chloroper-
benzoic acid in acetone, yields 69-76% (L. W. Deady, Synth.
Comm., 1977, 7, 509). This method can also be used for the
oxidation of phenylazopyridines (E. Buncel et al. , Canad. J.
Chem., 1984, 62, 1628) as well as for the N-oxidation of
2,2'-bipyridines (D. Wenkert and R. B. Woodward, J. org.
Chem., 1983, 48, 283). The same method can be applied to di-
N-oxides of 2,2':6',2"-terpyridine while the tri-N-oxide is
produced by hydrogen peroxide oxidation (R. P. Thummel and Y.
Jahng, J. org. Chem., 1985, 50, 3635). 2,6-Dideuteriopyridine
N-oxides can be prepared via the corresponding N-methoxypy-
ridinium chlorides (B. Nowak-Wydra and M. Szafran, Polish J.
Chem., 1981, 55, 1957).

(2) Deoxygenation

Many reducing agents can be used for the deoxygenation of a
variety of substituted pyridine N-oxides as pyridine N-
oxides themselves are mild oxidizing agents, such as catalytic
hydrogenation (Ni or Pd/c), hydrides (NaBH$_4$, LiAlH$_4$), phospho-
rus(III) compounds in the order of reactivity: PCl$_3$ > (PhO)$_3$P

---

*Reviews on pyridine N-oxide reactions can be found in; R. A.
Abramovitch and I. Shinkai, "Aromatic substitution via new rearrangement of heteroaromatic N-oxides", Acc. Chem. Res.,
1976, 9, 192; S. Oae and K. Ogino, "Rearrangements of t-Amine
Oxides", Heterocycles, 1977, 6, 583; R. A. Abramovitch, "Pyridine N-oxides", Lectures in Heterocyclic Chemistry, 1980, 5,
15.

> $(EtO)_3P \gg Ph_3P > Bu_3P$. Other suitable reagents are: $SOCl_2$, SCl_2, $TiCl_3$, $Na_2S_2O_4$, N_2H_4, Zn/AcOH, and Zn/NaOH. Chlorotrimethylsilane/sodium iodide/zinc, 45–95% (T. Morita et al., Chem. Letters, 1981, 921). Hexamethyldisilane/Bu_4NF, 51–92% (H. Vorbrüggen and K. Krolikiewicz, Tetrahedron Letters, 1983, 5337), Molybdenum(III)chloride via molybdenum(V)chloride/zinc in THF, ca. 50% (S. Polanc, B. Stanovnik, and M. Tisler, Synthesis, 1980, 129). Diphosphorus tetraiodide, 25–97% (H. Suzuki, N. Sato, and A. Osuka, Chem. Letters, 1980, 459). $TiCl_4/NaBH_4$ in dimethoxyethane, 95–97% (S. Kano, Y. Tanaka, and S. Hibono, Heterocycles, 1980, 14, 39). Reduction with sulphur monoxide generated from E-2,3-diphenylthiiran-1-oxide, 30–85% (B. F. Bonini et al., Tetrahedron Letters, 1979, 1799). $TiCl_3$ in methanol 97% (J. M. McCall and R. E. Ten Brink, Synthesis, 1975, 335. Chromium(II)chloride in acetone/water (Y. Akita et al., Chem. pharm. Bull. Japan, 1976, 24, 1839). Irradiation of 4-nitropyridine N-oxides in methylene chloride with trimethyl phosphite yields, 85–93%, the corresponding 4-nitropyridines (C. Kaneko, A. Yamamoto, and M. Gomi, Heterocycles, 1979, 12, 227). O-Arylation of pyridine N-oxides via N-(3-nitrophenoxy)pyridinium tetrafluoborates has been reported (R. A. Abramovitch and M. N. Inbasekaran, Tetrahedron Letters, 1977, 1109, and R. A. Abramovitch et al., J. org. Chem., 1979, 44, 464).

(3) Deoxidative substitution

As the N-oxide function activates the pyridine ring, a number of reagents will transform the pyridine N-oxides into ring substituted pyridines by deoxidative substitution after initial reaction at the oxygen atom. 5-Nitropyridine N-oxides can undergo thermal intramolecular hydroxylation (P. G. Sammes, G. Serra-Errante, and A. C. Tinker, J. chem. Soc. Perkin I, 1979, 1736).

36%   5%

Thermolyses of pyridine N-oxide-$SbCl_5$-complexes yields, 70–80%, the corresponding $2(1H)$-pyridones (J. Yamamoto et al., Tetrahedron, 1981, 37, 1871). The preparation of 2-chloromethylpyridine from the N-oxide with $POCl_3$ gives 2-chloromethylpyridine in 98% yield (M. L. Ash and R. G. Pews, J. heterocyclic Chem., 1981, <u>18</u>, 939), see also (J. H. Barnes *et al.*, Tetrahedron, 1982, <u>22</u>, 3277). Reaction of 4-cyanopyridine *N*-oxide with $POCl_3/PCl_5$ gives 3-chloro-4-cyanopyridine, 73% (J. Rokach and Y. Girard, J. heterocyclic Chem., 1978, <u>15</u>, 683). The cyanation of 3-halogeno-, 3-methoxy- and 3-dimethylamino *N*-oxide with trimethylsilane carbonitrile gives the corresponding 3-substituted 2-pyridine carbonitriles (T. Sakamoto *et al.*, Chem. pharm. Bull., 1985, <u>33</u>, 565).

Reaction of 4-substituted pyridine *N*-oxides with 3-arylrhodanines produces the (2-pyridyl)rhodanines (M. M. Yosif, S. Saeki, and M. Hamana, J. heterocyclic Chem., 1980, <u>17</u>, 305).

23-84%

A comprehensive study of the Katada reaction, the reaction of 2-picoline *N*-oxide with acetic anhydride to give 2-acetoxymethylpyridine has been reported (A. McKillop and M. K. Bhagrath, Heterocycles, 1985, <u>23</u>, 1697).

The thermal reaction of pyridine *N*-oxides and perfluoropropene yields, 20-60%, the corresponding 6-(1,2,2,2-tetrafluoroethyl)pyridine and carbonyl flouride (R. E. Banks, R. N. Hazeldine, and J. M. Robinson, J. chem. Soc. Perkin I, 1976, 1226).

Alkylthiopyridines can be prepared from pyridine *N*-oxide, an acyl chloride, and an alkylthiol (L. Bauer, T. E. Dickerhofe, and K. T. Tserng, J. heterocyclic Chem., 1975, <u>12</u>, 797). Tetrahydropyridines can also be obtained from pyridine *N*-oxides and alkyl mercaptans in acetic anhydride (S. Prachayasittikul, J. M. Kokosa, and L. Bauer, J. org. Chem., 1985, <u>50</u>, 997 and references cited therein). Flash vacuum pyrolyses (F. V. P.) of pyridine *N*-oxides give 2-picolyl-radicals resulting in annellated pyridines (A. Ohsawa, T. Kawaguchi, and Th. Igeta, J. org. Chem., 1982, <u>47</u>, 3497.

Diazotation of 2-(2-aminophenyl)pyridine *N*-oxide gives benzisoxazolo[2,3-*a*]pyridinium salts (R. A. Abramovitch and M. N.

Inbasekaran, Chem. Comm., 1978, 149).

Isocyanates and pyridine N-oxides yield interesting 1:1 adducts (R. A. Abramovitch, I. Shinkai, and R. V. Dahmm, J. heterocyclic Chem., 1976, <u>13</u>, 171).

2-Aminopyridine N-oxides react with thiophosgene to give $2H$-[1,2,3]oxadiazolo[2,3-a]-pyridin-2-thiones (D. Rousseau and A. Tauring, Canad. J. Chem., 1977, <u>55</u>, 3736), the products explode on heating.

A $2H$-azepine can be prepared by cycloaddition of a cyclo-alkyne and pyridine N-oxide (A. Krebs *et al.*, Heterocycles, 1979, <u>12</u>, 1153).

Useful reactive dienes can be prepared from 2-chloromethyl-pyridine N-oxide (M. Riediker and W. Graf, Chimia, 1980, <u>34</u>, 461).

Nucleophilic reagents derived from malonic acid derivatives react at the 3-position of 3-bromo-4-nitropyridine N-oxides (E. Matsumura and M. Ariga, Bull. chem. Soc. Japan, 1977, 50, 237). A number of interesting thermolytic ring contractions of 2-azidopyridine N-oxides have been reported (R. A. Abramovitch and B. W. Cue, J. Amer. chem. Soc., 1976, 98, 1478) while photolysis gives the 1,2-oxazines (R. A. Abramovitch and C. Dupuy, Chem. Comm., 1981, 36).

*(iii) Pyridinium ylides (betaines) and related zwitterions**

The reactive pyridiniumarylsulphonylmethylides undergo 1,3-dipolar cycloadditions (R. A. Abramovitch and V. Alexanian, J. org. Chem., 1976, 41, 2144).

---

*For reviews on the cycloaddition reactions of 3-oxidopyridinium ylides, see N. Dennis, A. R. Katritzky, and Y. Takeuchi, Angew. Chem., 1976, 88, 41; as well as A. R. Katritzky and N. Dennis in "New trends in heterocyclic chemistry", R. B. Mitra *et al.* ed., Studies in organic chemistry 3, Elsevier, Amsterdam, 1979, 290. For reviews on pyridinium N-ylides, see J. Streith, Pure appl. Chem., 1977, 49, 305; G. Surpateanu *et al.*, Tetrahedron, 1976, 32, 2647.

The pyridinium 4-toluenesulphonylmethylides can also serve as a formyl anion equivalent (R. A. Abramovitch *et al.*, Tetrahedron Letters, 1980, 705).

Heteroaromatic rings will also stabilize ylides useful in various Kröhnke reactions (A. R. Katritzky and D. Moderhack, J. chem. Soc. Perkin I, 1976, 909), such ylides are easily alkylated, acylated, and carbamoylated at the α-CH$_2$ group, while exchange at the pyridine C-2 and C-6 hydrogens is faster than exchange at the α-CH$_2$ hydrogens with electrophiles, such as benzaldehyde (A. R. Katritzky *et al.*, J. chem. Soc. Perkin I, 1984, 941).

Pyridinium ylides can undergo a number of fragmentation reactions, for example pyridinium diacylmethanides are cleaved by ozone yielding vicinal tricarbonyl compounds (K. Schank and C. Lick, Ber., 1982, 115, 3890).

3-Aminothien-4-ylpyridinium ylides can be prepared from 1-cyanomethylpyridinium salts, subsequent nucleophilic ring opening of the pyridinium ring yields 3,4-diaminothiophenes (Y. Tominaga *et al.*, Heterocycles, 1977, 6, 1871).

Interesting stable ylides can be obtained from 3,4-dioxocyclo-butenyl dichloride (squaric acid dichloride), (J. Grünefeld and G. Zinner, Chemiker Zeitung, 1984, 108, 112) and (A. H. Schmidt, A. Aimene, and M. Schneider, Synthesis, 1984, 436).

32-76%

A number of intramolecular cyclisations of pyridinium *N*-imines based on reaction either at the pyridine C-2 or at a C-2 substituent, such as a 2-methyl group, have been reported. An illustrative example is given below (A. Kakehi *et al.*, Bull. chem. Soc. Japan, 1978, 51, 251).

For other example see (A. Kakehi *et al.*, Chem. Letters, 1976, 413), (H. Ochi *et al.*, Bull. chem. Soc. Japan, 1976, 49, 1980), and (H. Fujito *et al.*, Heterocycles, 1977, 6, 379).

Pyridinium *N*-methylides can undergo 1,3-dipolar cycloaddi-tions with dipolarophiles, such as benzyne and olefins (K. Matsumoto *et al.*, Chem. Letters, 1982, 869) and (O. Tsuge *et al.*, *ibid.*, 1984, 465).

100%

A new type of ylide is the 3-imidopyridinium ylide prepared from 3-methylsulphonylaminopyridine (N. Dennis *et al.*, J. chem. Soc. Perkin I, 1977, 1304).

The ylide structure has been confirmed by an X-ray structure
determination. A large number of 3-oxidopyridinium ylides can
be prepared from 3-hydroxypyridines, for example (A. R.
Katritzky, B. A. Boyce, and N. Dennis, Polish. J. Chem., 1981,
55, 1351).

These and similar ylides react as 1,3-dipoles with a variety
of 2π electron addends (see the references to the reviews mentioned at the beginning of this section). Other examples are
given by A. H. Moustafa, A. A. Shalaby, and R. A. Jones, (J.
chem. Res., 1983, 37) in which the 3-oxidopyridinium ylide, R
= pyridazin-3-yl reacts with a number of π electron-excessive
and -deficient alkenes, in which the pyrazin group acts as a
leaving group.

Thus, 3-oxidopyridinium ylides are versatile cycloaddition
components, they will undergo cycloaddition at four different
orientations: (i) as 4π component across the 2,6-positions
(ii) as a $2\pi/6\pi$ component across the 2,4-positions (iii) as an
8π component either across the oxygen atom and at the 4-posi-
tion or (iv) across the oxygen atom and the 2-position of the
ring

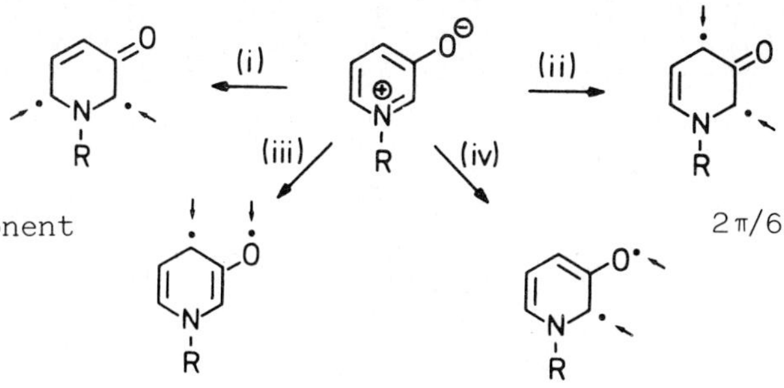

4π component

2π/6π component

8π component   8π component

The following types of products have been obtained.

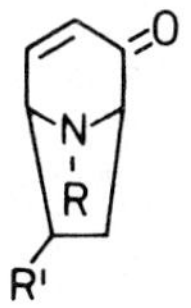

2π-thermal 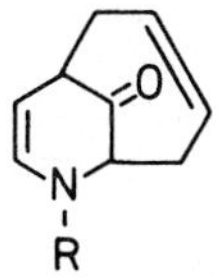   4π-thermal    6π-thermal

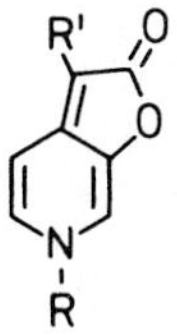

2π-thermal 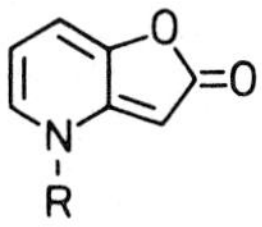   2π-thermal 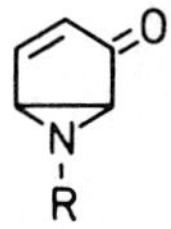   isomerisation-hν

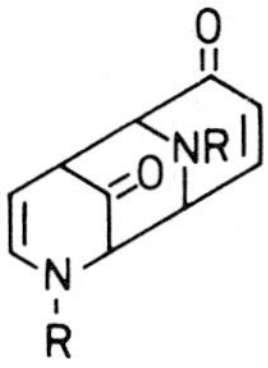

dimerisation-thermal 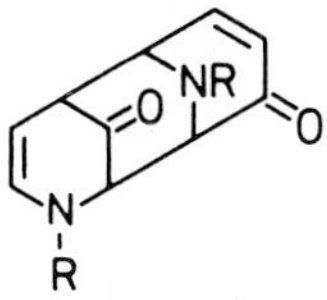 dimerisation-thermal 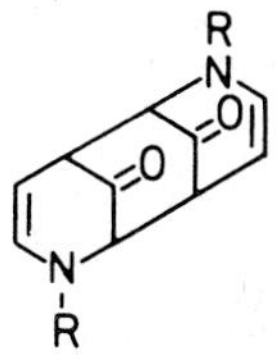 dimerisation-hν

Further examples can be found in the following paper: M. Hamaguchi, H. Matsuura, and T. Nagai, Chem. Comm., 1982, 262; G. Ferguson *et al.*, *ibid.*, 1983, 1216, and A. R. Katritzky *et al.*, J. chem. Soc. Perkin I, 1980, 1176 and references cited therein.

A general theory which relates the structure and reactivity

of dipolar heterocycles, such as 3-oxidopyridinium ylides, has been presented (C. A. Ramsden, Chem. Comm., 1977, 109).

Photooxidation of 3-oxidopyridinium ylides gives rise to maleimides (J. Banerji, N. Dennis, and A. R. Katritzky, Heterocycles, 1976, 5, 71) via intermediate hydroperoxide.

Hydrogen peroxide oxidation of related pyridinium ylides can give mixtures of arylpyrroles and 3-oxidopyridinium ylides via related intermediates (A. R. Katritzky et al., J. chem. Soc. Perkin I, 1980, 1870).

(b) Pyridines as benzene analogues: Substitution at the ring carbon atoms

(i) Electrophilic substitution

The influence of substituents in the pyridine nitramine rearrangement for a series of substituted aminopyridines has been reported (L. W. Deady, O. L. Korytsky, and J. E. Rowe, Austral. J. Chem., 1982, 35, 2025 and references cited therein). Halogenation of pyridines can be facilitated by formation of palladium(II)chloride complexes (S. Paraskewas, Synthesis, 1980, 378).

Pyridine reacts with xenon difluoride to give a mixture of 2-fluoro-, 3-fluoro-, and 2,6-difluoropyridine (S. P. Anand and R. Filler, J. fluorine Chem., 1976, 7, 179). Passing methylpyridines over caesiumtetrafluorocobaltate at high temperature yields polyfluoropyridines (R. G. Plevey, R. W. Rendell, and J. C. Tatlow, ibid., 1982, 21, 265).

3-Ethoxypyridine can be alkylated under Friedel-Crafts conditions with alkyl chlorides to give 2-benzyl-3-ethoxy- or 5-cyclohexyl-3-ethoxypyridine (F. M. Saidova and E. A.

Filatova, J. org. Chem. USSR, 1977, 13, 1231). However, it is well known that conventional electrophilic substitutions, such as Friedel–Crafts reactions, fail in the simple pyridines. It has been demonstrated that 3-acylpyridines can be obtained *via* indirect acylation of dihydropyridine species under Friedel–Crafts conditions (D. L. Comins and N. B. Mantlo, Tetrahedron Letters, 1983, 3683).

30–49% overall

Sulphene reacts with 2,6-dimethoxypyridine at C-3 (J. S. Grossert *et al.*, Chem. Comm., 1982, 1175).

80%

Direct ring benzylation of 2,6-diaminopyridine by fusion with benzyl chloride has been reported (W. Czuba and P. Kowalski, Polish J. Chem., 1981, 55, 931).

It is possible to aminoarylate nitropyridines *via* heterocyclic benzidine-like rearrangements (T. Sheradsky, E. Nov, and S. A. Grisaru, J. chem. Soc. Perkin I, 1979, 2902 and references cited therein). This reaction involves an initial nucleophilic displacement of a reactive halogen followed by an intramolecular electrophilic substitution.

An important reaction type is the directed metallation of pyridines, in which a lithiated pyridine reacts with an electrophile at low temperature.

Direct metallation of pyridines is only known in a few cases, mainly because organlithium reagents add across the C=N bond, however lithiated pyridines can be obtained by metal halogen exchange, as shown in the following example with 3-bromopyridine (W. E. Parham and R. M. Piccirilli, J. org. Chem., 1977, 42, 257).

For related *ortho*-lithiation of halogenopyridines and subsequent reactions with electrophiles see (M. Mallet and G. Queguiner, Tetrahedron, 1979, 35, 1625; G. W. Gribble and M. G. Saulnier, Tetrahedron Letters, 1980, 4137). It has recently been demonstrated that pyridines can be functionalized *via* direct metallation (J. Verbeek *et al.*, Chem. Comm., 1984, 257 and J. Verbeek and J. Brandsma, J. org. Chem., 1984, 49, 3857). A large number of polysubstituted pyridines have been prepared *via* directed metallation reaction of pyridines activated by a variety of *ortho*-substituents according to the general scheme

Representative examples may be found in the following references*. For R = $CONR_2$ see (M. Wanatabe and V. Sniekus, J. Amer. chem. Soc., 1980, 102, 1457). For R = CON^-R see (A. R. Katritzky, S. Rahimi-Rastgoo, and N. K. Ponkshe, Synthesis, 1981, 127). For R = 2-oxazolinyl see (A. I. Meyers and R. A. Gabel, J. org. Chem., 1982, 47, 2633). For R = SO_2NR_2 see (P. Breant, F. Marsais, and G. Queguiner, Synthesis, 1983, 822). For R = NCO_2-*t*-Bu see (Y. Tamura *et al.*, Chem. pharm. Bull.

---

*A review of such reactions can be found in the paper by M. A. J. Miah and V. Sniekus, J. org. Chem., 1985, 50, 5436, where the $-OCONEt_2$ group is used.

Japan, 1982, 30, 1257). For R = alkoxy see (F. Marsais, G. Le Nard, and G. Queguiner, Synthesis, 1982, 235). For R = CO(CF$_3$)$_2$ see (S. L. Taylor, D. Y. Lee, and J. C. Martin, J. org. Chem., 1983, 48, 4158). For R = -NHCOC(Me)$_3$ see (J. A. Turner, J. org. Chem., 1983, 48, 3401). A representative example for all of these reactions is the following (T. Gungor, F. Marsais, and G. Queguiner, Synthesis, 1982, 499).

3-Arylpyridines can be prepared via diethyl-(3-pyridyl)borane obtained from 3-lithiopyridine (M. Ishikura, M. Kamada, and M. Tarashima, Synthesis, 1984, 936).

5-Arylnicotinates can be obtained in a related reaction from 5-bromonicotinates (W. J. Thompson and J. Gardino, J. org. Chem., 1984, 49, 5237). The easy direct lithiation of 1-methyl-4(1H)-pyridone has recently been demonstrated (P. Patel and J. A. Joule, Chem. Comm., 1985, 1021).

(ii) Nucleophilic substitution*

Nucleophilic substitution is easy at C-2 and C-4 in pyridines, furthermore nucleophilic displacement of a good leaving group by addition-elimination occurs with leaving groups at C-2 and C-4. Corresponding N-substituted pyridinium compounds are very easily attacked by nucleophiles at C-2 or C-4, some of these reactions have already been mentioned in the pyridine

---

*Nucleophilic attack at pyridines can be classified in the types A-E : (A. Katritzky and E. Lunt, Tetrahedron, 1969, 25, 4291), types A-D, and (R. A. Abramovitch and G. M. Singer, J. heterocyclic Chem., Supplement, 1974, 14, 1) type E.

N-oxide section page **34** . Nucleophilic reactions of pyridinium compounds not followed by elimination will be mentioned in the section on the hydrogenated pyridine derivatives, see page **93** .

(iii) Typical nucleophilic substitutions

(1) Displacement of hydrogen from pyridine*

Normally ether free alkyl magnesium halides and alkyl lithium compounds react with pyridine in the absence of free metals to form 2-alkylpyridines while the reaction of an alkyl halide with the metal *in situ* in pyridine gives a 4-alkylpyridine (D. Bryce-Smith, P. J. Morris, and B. J. Wakefield, J. chem. Soc. Perkin I, 1976, 1977). Reaction of organo lithium reagents with pyridines gives N-lithio-1,2-dihydropyridines which subsequently can react with an electrophile to give 2,5-disubstituted pyridines. The σ-complex in such reactions has been isolated (C. S. Giam *et al.*, J. chem. Soc. Perkin I, 1979, 3082 and T. A. Ondus *et al.*, Can. J. Chem., 1978, <u>56</u>, 1913).

$$\text{Pyridine} + \text{RLi} \longrightarrow \underset{\underset{Li}{|}}{\text{N}}\text{-dihydropyridine-R,H} \xrightarrow{R'X} \text{R'-pyridine-R}$$

R and R^1 = alkyl, aryl   X = NCO

Lithium diisopropylamide (LDA) reacts with ethyl nicotinate and with ethyl isonicotinate to give 4-nicotinyl nicotinic acid and 2-isonicotinoyl isonicotinic acid, respectively (M. Ferles and A. Silhankova, Coll. Czech. chem. Comm., 1979, <u>44</u>, 3137). Introduction of an electron withdrawing substituent, such as a nitro group into the pyridine ring, increases the reactivity of the pyridine toward nucleophilic attack, for example 2-, 3-, and 4-nitropyridines react easily at the positions *ortho* or *para* to the nitro group with a sulphone carbanion to give nitropyridyl sulphones (M. Makoza, B. Chylinska, and B. Mudryk, Ann., 1984, 8).

---

*A review on the Chichibabin reaction has been published; A. F. Pozharskii, A. M. Simonov, and V. N. Doronkin, "Advances in the study of the Chichibabin reaction", Russian chem. Rev., 1978, <u>47</u>, 1042. For the reaction of various alkali-metal derivatives and pyridines, see E. M. Kaiser, J. D. Petty, and P. L. A. Knutson, Synthesis, 1977, 509.

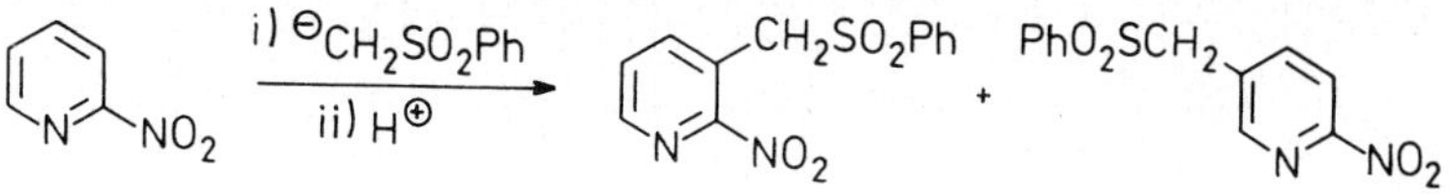

A direct nucleophilic hydroxylation of pyridine by heating it with anhydrous copper sulphate at 300° gives 2(1H)-pyridone in 95% yield (P. Tomasik and A. Woszcyk, Tetrahedron Letters, 1977, 2193). The nucleophilic displacement of hydrogen from a *pyridinium salt* usually takes place *via* an addition elimination process (AE reaction) and therefore involves a 1,2- or a 1,4-dihydropyridine derivative. For example carbanions derived from nitroalkane will react with N-(4-oxopyridinium-1-yl)pyridinium salts to give 1,4-dihydropyridines which upon irradiation result in the formation of 4-pyridyl-nitroalkanes (A. R. Katritzky *et al.*, J. chem. Soc. Perkin I, 1981, 588, as well as *ibid.*, 1981, 2476 and references cited therein, see also A. R. Katritzky *et al.*, Angew. Chem., 1979, 91, 856).

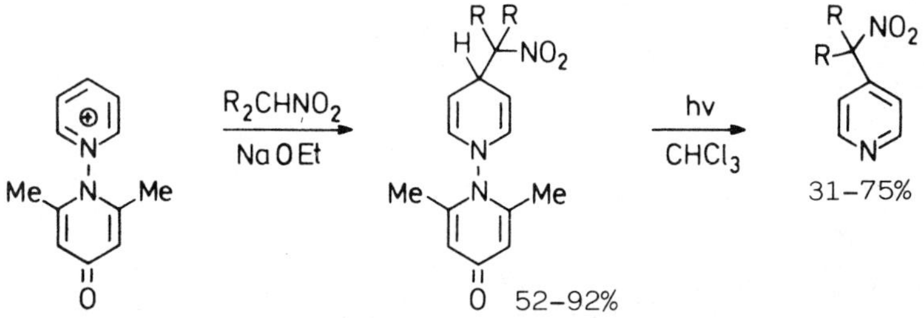

A nucleophile, such as indole, can react with a pyridinium compound, as the following example demonstrates (T. V. Stupnikova and V. V. Petrenko, Chemistry of heterocyclic compounds USSR, 1984, 1383).

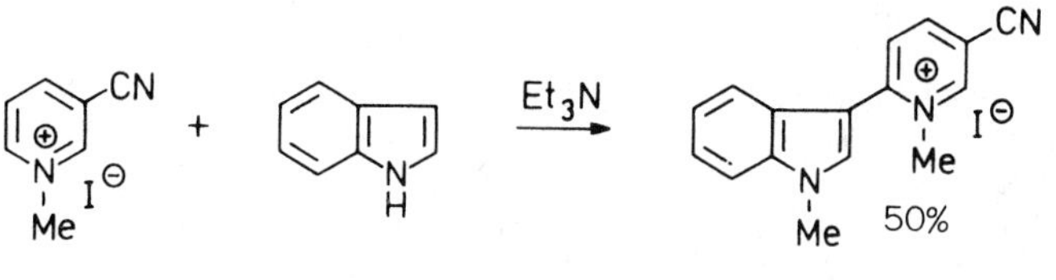

Grignard reagents often react regioselectively with 3-substituted pyridinium compounds. The following reaction provides a good example (D. L. Comins, R. K. Smith, and E. D. Stroud, Heterocycles, 1984, 22, 339).

X = Cl, Br, and 1,3-dioxolan-2-yl; R = alkyl, aryl

Other examples can be found in (R. E. Lyle and P. L. Comins, J. org. Chem., 1976, 41, 3250; D. L. Comins and N. B. Mantlo, J. heterocyclic Chem., 1983, 20, 1239 and in D. L. Comins and N. B. Mantlo, J. org. Chem., 1985, 50, 4410).

A pyridinium group is an excellent leaving group in reactions of the type.

This provides a method for the preparation of pyridine-4-phosphonic acids (B. Boduszek and J. S. Wieczorek, Synthesis, 1979, 452; B. Boduszek and J. S. Wieczorek, Monatsh., 1980, 111, 1111). Other applications of this method can be found in (A. R. Katritzky et al., J. chem. Soc. Perkin I, 1983, 2617), see also (A. R. Katritzky et al., Angew. Chem., loc. cit.). 1-Triphenylmethylpyridinium salts give pyridyl-4-phosphonates in a regioselective reaction with sodio phosphonates, with pyridinium salts unsubstituted at the 2,6-positions the resulting pyridine phosphonate is formed from the 1,4-intermediate by elimination of triphenylmethane (D. Redmore, J. org. Chem., 1976, 41, 2148). An interesting regiospecific one-pot synthesis of 4-substituted pyridines can be carried out via alkylation of diisopropyl 1-ethoxycarbonyl-1,4-dihydropyridine-4-phosphonate and subsequent treatment with butyllithium (K. Akiba, H. Matsuoka, and M. Wada, Tetrahedron Letters, 1981, 4093; K. Akiba, Y. Iseki, and M. Wada, ibid., 1982, 429). 1-Ethoxycarbonylpyridinium chloride reacts with a number of α-functional organometallics to give 1,2- or 1,4-dihydropyridines. Subsequent oxidation yields the substituted pyridines (G. Courtois, A. Al-Arnaout, and L. Miginiac, Tetrahedron Letters, 1985, 1025, this paper also gives a review on the relevant literature).

In a regioselective nucleophilic reaction with organolithium reagents stable 1,4-dihydropyridines can be formed (C. S. Giam and A. E. Hauch, Chem. Comm., 1978, 615).

R = alkyl, aryl   ∼63%

The reaction of Grignard reagents and pyridine *N*-oxide may result in ring opened products and often reactions of this type are not regiospecific, however a regiospecific one pot synthesis of 2-substituted pyridines from pyridine *N*-oxide has been described (T. R. Webb, Tetrahedron Letters, 1985, 3191).

R' = aryl, vinyl, and $-C\equiv CSiMe_3$

(2) Displacement of halogen

Nucleophilic displacement of halogen usually occurs by an AE or an EA mechanism. The reactivity of 2-halogens in pyridinium compounds has been exploited for a range of carboxylic acids, alcohols etc. (a review can be found in: T. Mukaiyama, Angew. Chem., 1979, 91, 798). A typical reaction is given below.

X = F, Cl, Br or I

Related applications of this method can be found in (M. Furukawa *et al.*, Synthesis, 1978, 441 and in A. J. S. Duggan, E. J. J. Grabowski, and W. K. Russ, *ibid.*, 1980, 573).

2-Chloro- and 6-chloro-3-trichloromethylpyridine behaves as ambident electrophilic substrates towards methoxide which re-acts at the 2-position as well as at the trichloromethyl group (R. S. Dainter, H. Suschitzky, and B. Wakefield, Tetrahedron Letters, 1984, 5693). The palladium catalyzed reaction of

3-iodopyridine (K. Edo, T. Sakamoto, and H. Yamanaka, Chem. pharm. Bull. Japan, 1979, 27, 193) and of 4-bromopyridine derivatives (Y. Tamaru *et al.*, Chemistry Letters, 1978, 975) with alkenes has been reported to give 3- and 4-alkylpyridines, respectively.

(2,6)-Pyridinophanes can be prepared in a single step by nickel catalyzed Grignard cyclocoupling (K. Tamao *et al.*, J. Amer. chem. Soc., 1975, 97, 4405).

n = 6-12,

10-33%

Pyridyldiphenylphosphines can be prepared from halopyridines and lithium diphenylphosphide (G. R. Newcome and D. C. Hayer, J. org. Chem., 1978, 43, 947). Various enolate anions readily displace chloro, pyridyl, and acetoxy substituents from the 4-position of simple pyridine derivatives (F. X. Smith and G. G. Evans, J. heterocyclic Chem., 1976, 13, 1025) under reaction conditions where the pyridine ring carries a positive charge, solvent $DMF/(MeCO)_2O$.

89%

It is possible to prepare 2-, 3-, and 4-iodopyridine from the corresponding chloro- or bromopyridines *via* the trimethylstannyl derivatives in fair yields (Y. Yamamoto and A. Yanagi, Heterocycles, 1981, 16, 1161). Alkylthiopyridines can be prepared from sodium thiolates, procedures have been reported for a number of substituted pyridines (S. G. Woods *et al.*, J. heterocyclic Chem., 1984, 21, 97 and L. Testaferri *et al.*, Tetrahedron, 1985, 41, 1373).

88%

(3) Displacement of nitrogen containing groups

The displacement of a nitro group is a well known procedure for preparation of alkylthio aryls, thus 5-alkylthiopyridines can be prepared by this method (N. Finch *et al.*, J. med. Chem., 1978, 21, 1269).

O_2N-[pyridine ring]-CO_2Me → (1) NaSR (2) $OH^{\ominus}$ → RS-[pyridine ring]-CO_2H

(4) Displacement of oxygen and sulphur containing groups

Rate parameters for the nucleophilic aromatic substitution of 4-methoxypyridinium compounds follow general base catalysis similar to nitroactivated benzene systems (R. Aucta, G. Doddi, and G. Illuminati, J. Amer. chem. Soc., 1983, 105, 5661). Both sulphinyl and sulphonyl groups in the pyridine, either at the 2- or the 4-position can be displaced with nucleophiles, such as RO^- or RS^- (N. Furukawa *et al.*, Tetrahedron Letters, 1983, 3243) in the sequential order $RSO_2 \simeq RSO > Br \simeq Cl \gg RS$.

A 4-sulphonylmethyl group is more reactive than a 2- or 6-sulphonylmethyl group (S. D. Moshchitskii, G. A. Zalesskii, and V. P. Kukhar, Chem. heterocyclic Compounds, USSR, 1976, 915).

Sulphide groups are not readily displaced by nucleophiles, but examples of this type of reaction may be useful in some cases, thus 2,6-bis(methylthio)pyridine reacts with sodium methoxide to yield 2-methoxy-6-methylthiopyridine (L Testaferri *et al.*, Tetrahedron, 1985, 41, 1373). See also page 83 .

(v) Radical substitution

Radical substitutions in the pyridine ring can be initiated either thermally by suitable radical initiaters or photochemically. The radical photochemical substitutions will be covered in section *(vi)* of this chapter.

(1) Alkylation*

Cyclohexyl radicals, generated by thermal decomposition of *t*-butylperoxyoxalate in cyclohexane, react with 2- and 3-X-pyridines (X = CN, COMe, and CO_2Me) to give a single substitution product by reaction at the pyridine 5- and 6-position, respectively, yields 35–50% (D. Chianelli *et al.*, Tetrahedron, 1982, 38, 657). Radicals can also be generated by γ-irradiation, thus a simple homolytic substitution reaction for methylation of methylnicotinate has been reported (G. R. Newkome and C. R. Marston, J. org. Chem., 1985, 50, 4162), a discussion of related reactions can also be found in this paper.

Acyl groups can be displaced by 1-adamantyl and other alkyl radicals in pyridine radical cations (M. Fiorentino *et al.*, Chem. Comm., 1976, 329).

(2) Arylation**

Arylation of pyridine with phenyl radicals follows the positional reactivity 2 > 3 >> 4 while a protonated pyridine shows the reactivity 4 > 2. The following general mechanism for a S_R reaction on a protonated pyridine can be formulated as

---

*For a review covering some aspects of alkylation reactions see F. Minisci, "Recent aspects of homolytic aromatic substitutions", Topics in Current Chem., 1976, 62, 1.
**For a review see G. Vernin, "Recent progress in heteroarylation reactions", Bull. chim. Soc. France., 1976, 1257.

(3) Acylation

Carboxamido groups are introduced into the 6-position of 3-acylpyridines with high regioselectivity *via* the Minisci reaction with formamides, Fe(II)-sulphate and *t*-butyl hydroperoxide (E. Langhals, H. Langhals, and C. Rüchardt, Ann., 1982, 930, see also E. Langhals and H. Langhals, and C. Rüchardt, Chem. Ber., 1984, 117, 1259). A typical reaction is given below.

(4) Other radical substitutions

It has been found that the relative reactivity of 2-halopyridines towards the benzenethiolate anion is: 2-iodopyridine ≃ 2-bromopyridine >> 2-chloropyridine ≃ 2-fluoropyridine. When the reaction is carried out in DMF, this type of reaction takes place *via* a radical chain process (S. Kondo, M. Nakanishi, and T. Tsuda, J. heterocyclic Chem., 1984, 21, 1243). Fluorination of pyridine at high temp. with caesium tetrafluorocobaltate(III) yields a mixture of polyfluoropyridines *via* a pyridine radical cation (R. G. Plevey, R. W. Rendell, and J. C. Tatlow, J. fluorine Chem., 1982, 21, 159).

(vi) Photochemical reactions of pyridine and its derivatives*

The photochemistry of pyridine *N*-oxide, pyridine betaines, and related compounds has been a rapidly developing field of research. But in contrast, when compared with benzene, pyridine has a rather sparse photochemistry.

---

*For a review see A, Lablanche-Combier, "Photochemistry of Heterocyclic compounds" Chapter 4, O. Buchardt ed., Wiley, New York, 1976.

(1, 2, and 3) Reactions of substituents, addition and substitution at the ring carbon atoms

Halogenated pyridines are versatile compounds in this connection, thus 3-chlorotetrafluoropyridine reacts photochemically with ethylene or cycloalkanes by insertion into the C-Cl bond, yielding 3-alkyltetrafluoropyridines in good yields (M. G. Barlow, R. N. Hazeldine, and J. R. Langridge, Chem. Comm., 1979, 608). However, pentafluoropyridine gives 1:1- and 2:1-adducts when irradiated with ethylene (M. G. Barlow, D. E. Brown, and R. N. Hazeldine, J. chem. Soc. Perkin I, 1978, 363).

37%   21%

Photochemical dehalogenation and arylation of pentachloro- and pentabromopyridine takes place regioselectively leading to loss of the 3-halogen (J. Brat *et al.*, J. chem. Soc. Perkin I, 1980, 648).

40%

3-Bromopyridine reacts with ketone enolates, such as potassio-acetone in a photochemical $S_{RN}1$ reaction (A. P. Komin and J. F. Wolfe, J. org. Chem., 1977, 42, 2481).

95%

Pyridine can react photochemically with alkylamines, such as diethyl- and triethylamine at the amine α-CH$_2$-carbons (A. Gilbert and S. Krestonosich, J. chem. Soc. Perkin I, 1980, 2531) to give 2- and 4-substituted pyridines. Irradiation of pyridine-2-carbonitrile in acidic solution gives 6-ethoxy-pyridine-2-carbonitrile and 2-(1-hydroxyethyl)pyridine (T. Furihata and A. Sugimori, Chem. Comm., 1975, 241). Protonated 4-cyanopyridine undergoes a photochemical displacement reaction when irradiated in isopropanol with benzophenone yielding a mixture of diphenyl-(4-pyridyl)carbinol and 4-benzhydryl-pyridine (B. M. Vittimberga, F. Minisci, and S. Morrocchi, J. Amer. chem. Soc., 1975, 97, 4397). Pyridine carboxylic esters can be photoalkylated or photoalkoxylated by irradiation in methanol or ethanol (T. Sugiyama et al., Chem. Letters, 1980, 131). An example is the following

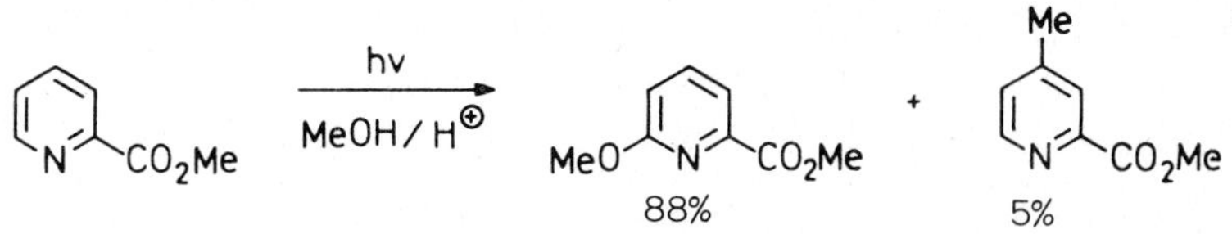

88%   5%

Photoalkylation of pyridine with dibromo N-methylmaleimide in acetone yields 3-substituted pyridines (K. M. Wald et al., Ber., 1980, 113, 2884).

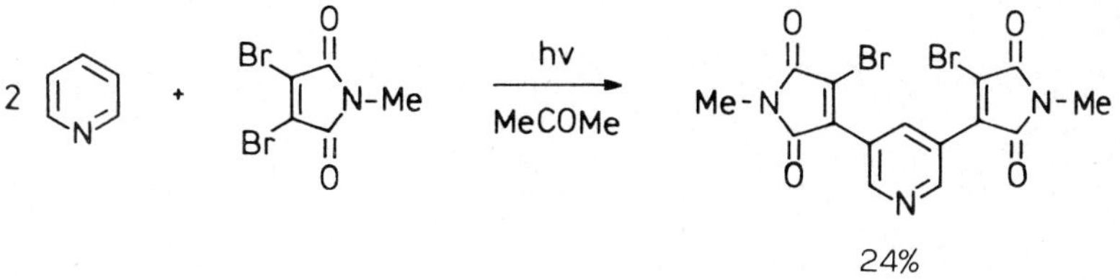

24%

Nitrile groups in pyridine can be displaced photochemically, thus irradiation of 2- and 4-pyridinecarbonitrile with cyclopentene in acetone gives the corresponding 2-cyclopentylpyridines (R. Bernardi et al., Tetrahedron Letters, 1981, 115).

Direct arylation of pyridine at the 3-position is difficult, an important regiospecific photoarylation reaction for the preparation of 3-phenylpyridines in fair yields has been reported (F. S. Tanaka, R. G. Wien, and B. L. Hoffer, Synth. Comm., 1983, 13, 951).

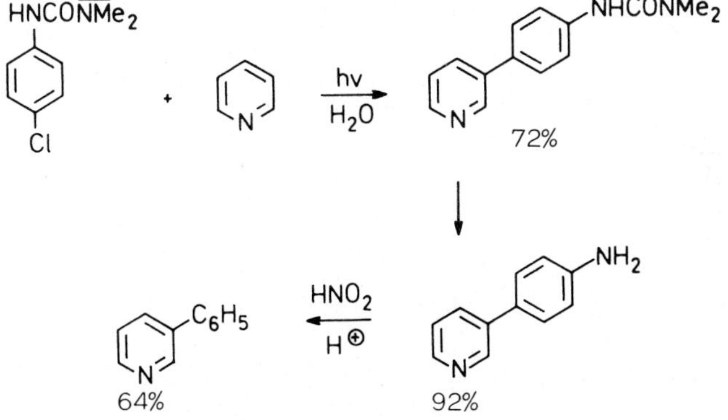

72%

64%   92%

(4) Rearrangements

Irradiation of alkyl 2-pyridylacetate gives a new Dewar pyridine (Y. Ogata and K. Takagi, J. org. Chem., 1978, <u>43</u>, 944).

Also azaprismanes can be obtained from highly fluorinated
pyridines (R. D. Chambers and R. Middleton, J. chem. Soc.
Perkin I, 1977, 1500).

Irradiation of 3- and 4-azidopyridines under basic conditions
gives monocyclic 1,3- and 1,4-diazepines (H. Sawanishi *et al.*,
Chem. pharm. Bull. Tokyo, 1984, <u>32</u>, 4694).

(5) Photochemical reactions of pyridine N-oxides and related compounds*

Irradiation of pyridine N-oxides under neutral or acidic conditions usually results in complicated reaction mixtures obtained with poor material balance. However, irradiation of the unsubstituted pyridine N-oxide in aqueous base quantitatively yields an enol anion by ring opening of the N-C bond (L. Finsen et al., Acta Chem. Scand., 1980, B34, 513).

In the presence of secondary amines the corresponding conjugated nitriles are obtained (J. Becher et al., Tetrahedron, 1981, 38, 789). The enol anion which is formed is stable due to delocalisation of the same type as found in the stable glutaconaldehyde anion whereas the protonated form

rapidly polymerizes leading to intractable tars of the type found when pyridine N-oxide is irradiated in the absence of base.

Irradiation of pyridine N-oxide in the presence of copper(II) ions gives rise to 40% of 2-formylpyrrole via a nitrene intermediate (F. Bellamy and J. Streith, J. chem. Res. (S), 1979, 18). Photolysis of 2-azidopyridine N-oxide results in the formation of a 1,2-oxazepine which is further thermally rearranged to a pyrrole derivative (R. A. Abramovitch and C.

---

*A review of the literature up to 1980 on pyridine N-oxide photochemistry can be found in O. Buchardt et al., Acta chem. Scand., 1980, B34, 31. For other reviews on pyridine N-oxide photochemistry see A. Padwa, "Photochemistry of the carbon-nitrogen double bond", Chem. Rev., 1977, 77, 63; F. Bellamy and J. Streith, "The photochemistry of aromatic N-oxides, a critical review", Heterocycles, 1976, 4, 1391, in F. Bellamy, J. Streith, and H. Fritz, "Photochemistry of aromatic N-oxides", Nouveau Journal de Chem., 1979, 3, 115 and in a comprehensive review by A. Albini and M. Alpegiani, "Photochemistry of the N-oxide function, Chem. Rev., 1984, 84, 43.

Dupuy, Chem. Comm., 1981, 36).

(6) Photochemistry of pyridones

Usually 2(1*H*)-pyridones dimerise upon irradiation, however substituted 2(1*H*)-pyridones react photochemically by valence tautomerisation (C. Kaneko *et al.*, Chem. Comm., 1980, 1175; W. J. Begley *et al.*, J. chem. Soc., Perkin I, 1981, 2620, as well as references cited therein), for example

70%

The bicyclic compounds obtained in this reaction are useful for the preparation of functionalised β-lactams (J. Brennan, Chem. Comm., 1981, 880). The photocyclisation of *N*-alkyl-2(1*H*)-pyridones as well as the photocyclisation of *N,N*-polymethylene-bis-2(1*H*)-pyridones have been extensively studied (Y. Nakamura, J. Zsindely, and H. Schmid, Helv., 1976, 59, 2841; Y. Nakamura, T. Kato, and Y. Morita, Chem. Comm., 1978, 620; Y. Nakamura, T. Kato, and Y. Morita, *ibid.*, 1982, 1187 and references cited therein). Both (2+2)-, (4+2)-, and (4+4)-addition products have been isolated in these reactions.

51% (4+4)-addition

The main product was depicted as the *trans-anti* isomer, but the *trans-syn*; the *cis-anti*, and the *cis-syn* isomers are also isolated albeit in low yields.

If 1-methyl-2-pyridone is irradiated in the presence of 2-methyl-s-triazolo[1,5-a]pyridine, a (4+4)-cycloaddition takes place with formation of a photodimer incorporating the two different pyridine rings (T. Nagano, M. Hirobe, and T. Okamoto, Tetrahedron Letters, 1977, 3891).

1-Methyl-2($1H$)-pyridone can also form photoadducts with ethylenes forming bicyclic cyclobutane systems (K. Somekawa *et al.*, Bull. chem. Soc. Japan, 1981, 54, 1112; C. Kaneko, Chem. pharm. Bull. Japan, 1983, 31, 2168.

(vii) Non-photochemical ring fission, ring expansion, contraction, and rearrangement*

As pointed out at the beginning of this section of the 2nd edition, pyridinium compounds are readily ring opened by a number of nucleophiles, the most important being the hydroxyl ion. The general scheme for such reactions is (a type D nucleophilic reaction, see footnote on page 45).

Hydrolyses of the imine c or d results in the formation of a 2,4-pentadienal derivative e. But reactions of this type are in general dependent upon the type of substituent on the pyridine nitrogen, on other substituents in the pyridine ring as well as on the type of nucleophile used.

Recently a number of useful ring opening and ring closure reactions have been reported, including interesting intramolecular examples.

The primary reaction product b from a pyridinium compound and a nucleophile, when X = OH$^-$, is called a "pseudo-base", while the primary ring opened product c can be called a "non-

---

*Various aspects of pyridine ring rearrangements etc. have been described in the following reviews; V. Simanek and V. Preininger, "Pseudo-base formation from quaternary pyridinium quinolinium and isoquinolinium cations", Heterocycles, 1977, 6, 475; J. Becher, "Synthesis and Reactions of Glutaconaldehyde and 5-Amino-2,4-pentadienals", Synthesis, 1980, 589; A. N. Kost, S. P. Gromov, and R. S. Sagitullin, "Pyridine Ring Nucleophilic Recyclizations", Tetrahedron, 1981, 37, 3423; H. C. van der Plas, "Ring Degenerate Transformation of Azines", Tetrahedron, 1985, 41, 237.

cyclic pseudo-base". Usually *c* rearranges to the thermally more stable *all-trans* isomer *d*. Such compounds *b* are assumed to be intermediates in nucleophilic ring opening reaction of pyridinium compounds (reaction at the pyridine C-4 does not lead to ring opening). Stable examples of pseudo-bases have been characterised (W. H. Gündel, Ann., 1980, 1350).

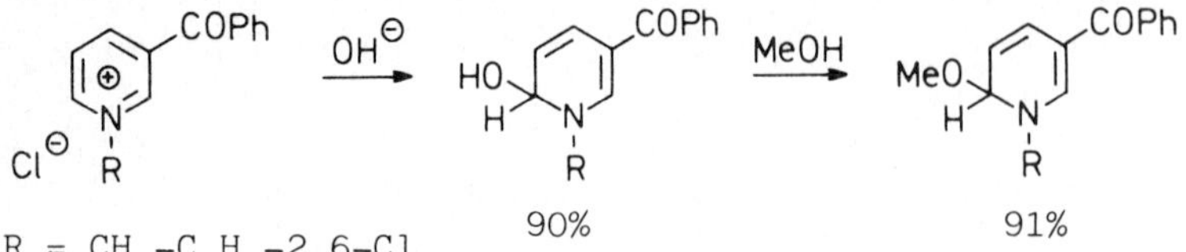

R = CH$_2$-C$_6$H$_3$-2,6-Cl$_2$

This type of reactions are often regiospecific, for example at -40°C 1,3-disubstituted pyridinium ions react with ammonia resulting in addition at C-6 when the C-3 group is CONH$_2$, CO$_2$Me, CF$_3$ or COMe, at C-2 when the C-3 group is Cl or I while a mixture is formed with CN as the C-3 group (J. A. Zoltewicz, L. S. Helmick, and J. K. O'Halloran, J. org. Chem., 1976, 41, 1303). The reactivity and pseudo-base formation from 3-cyano-1-methylpyridinium iodide has been discussed (F. M. Moracci *et al.*, Tetrahedron, 1976, 35, 2591). By the following reaction sequence it it possible to prepare the important parent 5-amino-2,4-pentadienal in 62% yield (D. Reinehr and T. Winckler, Angew. Chem., 1981, 93, 911).

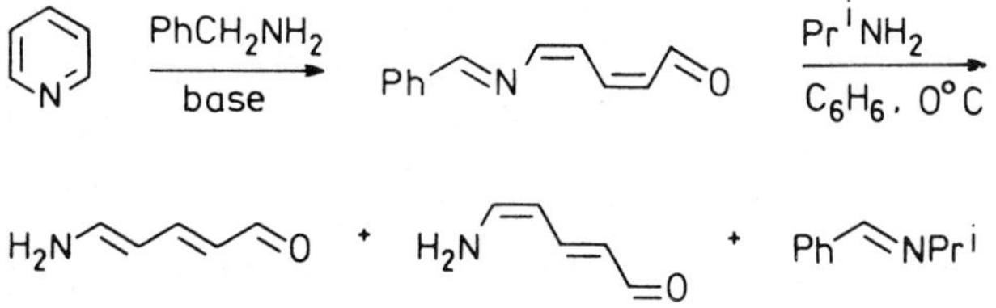

A preparatively useful procedure for salts of the glutaconaldehyde anion has also been reported (J. Becher, Org. Synth., 1980, 59, 79).

Starting from 4,4'-bipyridyl and using the classical pyridine ring opening reaction *via* 2,4-dinitrochlorobenzene it is possible to prepare 6,6'-biazulenyl (M. Hanke and C. Jutz, Synthesis, 1980, 31). Furthermore, from a pyridiniophanium salt it is possible to prepare large cyclodienes (H. Weber, J. Pant, and H. Wunderlich, Ber., 1985, 118, 4259).

A polymer, containing glutaconaldehyde groups, may be prepared *via* the cyanogen bromide ring opening of a pyridine polymer starting from 4-vinylpyridine (F. Pittner *et al.*, J. Amer. chem. Soc., 1980, 102, 2451).

Such polymers are useful for binding of proteins. Many 5-amino-2,4-pentadienal derivatives have been prepared from pyridines, see (K. Kigasawa *et al.*, Heterocycles, 1976, 4, 1257; H. Barth, M. Kobayashi, and H. Musso, Helv., 1979, 62, 1231; H. Takayama and T. Okamoto, Chem. pharm. Bull. Japan, 1978, 26, 2422), a typical example is given below.

A number of *N*-methoxypyridinium salts can be ring opened (J. Schnekenburger, D. Heber, and E. H. Brunschweiger, Arch. Pharm. Weinheim, 1982, 315, 817, see references cited for further examples of this type).

For the mechanism and stereochemistry of related ring opening reactions see (H. Sliwa and A. Tartar, Tetrahedron, 1979, 35, 341). As mentioned at the beginning of this section, ammonia can also be used as the nucleophile (J. A. Zoltewicz, L. S. Helmich, and J. K. O'Halloran *loc. cit.*).

The cyanide anion can be used as the nucleophile (C. W. F. Leuny, M. P. Sammes, and A. R. Katritzky, J. chem. Soc. Perkin I, 1979, 1698 as well as M. P. Sammes et al., ibid., 1983, 973). In this case the reaction takes place at the pyridine C-2.

Related ring opening products are described in the following references: R. D. Chambers, W. K. R. Musgrave, and P. G. Urben, Chem. Ind., 1975, 89; J. P. Scovill and J. V. Silverton, J. org. Chem., 1980, 45, 4372; T. Uno, K. Okumura, and Y. Kuroda, ibid., 1981, 46, 3175.

It has been demonstrated that cation substituted pyridinium compounds can be readily ring opened (G. Maas and B. Feith, Angew. Chem., 1985, 97, 518).

New examples of the use of thiophosgene/barium carbonate for the ring opening of pyridines have also been reported (R. Hull et al., Chem. Comm., 1983, 74).

2-Amino-1-alkylpyridinium salts give pentadienonitriles in

high yields (A. R. Katritzky, D. Winwood, and N. E. Grzeskowiak, Tetrahedron, 1982, $\underline{38}$, 1169).

Reaction of 2–bromo–6–lithiopyridine with trialkylboranes gives ring opened products upon acylation (K. Utimoto *et al.*, Tetrahedron, 1976, $\underline{32}$, 769).

If 2–bromo–3–lithiopyridine is treated with an excess of LDA (lithium diisopropylamide) followed by trimethylsilyl chloride, a related ring opening takes place (F. Marsais *et al.*, J. chem. Res. (S), 1982, 278).

Many ANROC (aromatic nucleophilic ring opening and ring closure) reactions have been reported in the pyridine series. As pyridinium salts are often easily ring opened, *vide supra* , intramolecular recyclizations of this type constitute an interesting preparative method, starting from a versatile suitable pyridine.

Nitropyridinium salts react with primary and secondary amines yielding *N*-alkylanilines as a result of recyclization of the ring opened pyridine ring (R. S. Sagitullin, S. P. Gromov, and A. N. Kost, Tetrahedron, 1978, $\underline{34}$, 2213 as well as the review by these authors cited at the beginning of this section), for example

19–30%

For other examples of this reaction type see (A. N. Kost, R. S. Sagitullin, and A. A. Fadda, Org. Prep. and Proc. Int., 1981, 13, 203; T. V. Stupnikova et al., Chem. heterocyclic Comp. USSR, 1983, 201; C. Ducrocq et al., J. chem. Soc. Perkin I, 1979, 135).

2-Pyridones may also undergo ring opening/ring closure reactions (M. Ariga, Y. Tohda, and E. Matsumura, Bull. chem. Soc. Japan, 1985, 58, 393), (not all intermediates have been shown in the schemes below as they can be found in the references).

By an ANROC reaction indoles can be prepared from nicotyrine using aqueous base (A. N. Kost et al., Chem. heterocyclic Comp. USSR, 1978, 1278).

Indoloquinolizidines can be obtained from pyridinium salts (M. Lounasmaa and T. Ranta, Heterocycles, 1983, 20, 1).

Quaternary salts of 3-pyridine carboximidamide give the 2-amino-3-pyridine carbaldehyde in 70% yield. This procedure is the best method for the preparation of this important starting material (W. H. Gündel, Z. Naturforsch., 1979, 34b, 1019; for related reactions see W. H. Gündel, ibid., 1980, 356, 896).

By ^{15}N-labelling it has been demonstrated that amination of halogenopyridines takes place via an ANROC mechanism (D. A. de Bie, B. Geurtsen, and H. C. van der Plas, J. org. Chem., 1985, 50, 484). Likewise the reaction of 2-halo-5-nitropyridines with the hydroxide ion takes place via ring opening and formation of an anion a of the same type as formed in the photochemical ring opening of pyridine N-oxide, see page 57 . (J. D. Reinheimer, L. L. Mayle, and G. G. Dolnikowski, J. org. Chem., 1980, 45, 3097).

Treatment of 2-(ethoxycarbonyl)pyridinium salts with primary amines results in an ANROC reaction (A. R. Katritzky, R. Awartani, and R. C. Patel, J. org. Chem., 1982, 47, 498).

N-Vinylpyridinium salts give interesting bicyclic compounds by ring opening with the hydroxyl ion (G. Palenik et al.,

Heterocycles, 1984, _22_, 717).

A related type of ring opening takes place with hydroxylamine as the nucleophile in which the intermediate glutaconaldehyde derivative also reacts intramolecularly in this case to a 1,2-oxazoline (A. Lorenzo and P. Molina, Ann. Quimica ser. C., 1981, _77_, 351).

Oxidation of phenylpyridinium salts leads to pyrroles (P. Nesvadba and J. Kuthan, Tetrahedron Letters, 1980, 3727).

For related reactions see A. R. Katritzky _et al._, Chem. Comm., 1979, 363.

Pyridazines as well as pyrazoles can be prepared by intramolecular ring opening and closure under Wolf-Kischner reaction conditions (M. M. Baradarani and J. A. Joule, J. chem. Soc. Perkin I, 1980, 72), intermediates not shown.

4. Individual compounds*

(i) Pyridine

The efficiency of various agents for the drying of pyridine and alkylpyridines has been compared (D. R. Burfield, R. H. Smithers, and A. S. C. Tan, J. org. Chem., 1981, <u>46</u>, 629). It is concluded that CaH_2 and CaC_2 as well as molecular sieves are the most effective desiccants for pyridine and alkyl-pyridines.

The chemistry of 1-substituted pyridinium compounds has been described on page **30** . It has been reported that pyridine borane C_5H_5N,BH_3 readily prepared from pyridine hydrochloride and sodium borohydride selectively reduces oximes to the corresponding hydroxylamines in good yields (Y. Kikugawa and M. Kawase, Chem. Letters, 1977, 1279).

_____

*For ^{15}N-NMR spectra of substituted pyridines, see H. J. Jakobsen, P. I. Yang, and W. S. Brey, Org. mag. Res., 1981, <u>17</u>, 290; ^{13}C-NMR shift increments of 3-pyridines, see R. Domisse et al., Heterocycles, 1981, <u>16</u>, 1893; Calculations of spin-spin coupling constants in pyridines, see S. A. T. Long and J. D. Memory, J. mag. Resonance, 1981, <u>44</u>, 355. A review on the advances in the chemistry of 2,6-disubstituted pyridines giving an access to the Russian literature has been published (L. N. Yakhontov and D. M. Krasnokutskaya, Russ. chem. Rev., 1981, <u>50</u>, 565).

(ii) Pyridine homologues: alkylpyridines*

Methyl groups in the 2,4- and 6-positions of pyridine rings can be converted into isopropyl (2- and 4-) and t-butyl (4-) groups by alkylation of the N-methylpyridinium iodides with methyl iodide (L. S. Hart et al., Chem. Comm., 1979, 24).

2- and 4-methylpyridines can be converted into the corresponding trichloromethylpyridines with $PCl_5/POCl_3$ (T. Kato, N. Katagiri, and A. Wagai, Tetrahedron, 1978, 34, 3445). Free radical chlorination of isopropylpyridines with t-butylhypochlorite affords the corresponding α-chloroalkylpyridines (H. Feuer and J. K. Doty, J. heterocyclic Chem., 1978, 15, 1517). Useful 5-trifluoromethylpyridines can be prepared from 5-methylpyridines (T. Haga et al., Heterocycles, 1984, 22, 117).

A systematic study of the selective side chain chlorination of 2,6-dimethylpyridine has been reported (Y. S. Harpman et al., Chem. heterocyclic Comp. USSR, 1980, 16, 89).

The easy lithiation of alkylpyridines is well known (see page 154 of the 2nd edition). A new synthesis of α-(2-pyridyl)ketones by acylation of 2-pyridylmethyl-lithium has been reported (R. P. Cassity, L. T. Taylor, and J. F.Wolfe, J. org. Chem., 1978, 43, 2286). The side chain lithiation of a number of pyrido[6]cycloalkanes has been described (J. Epsztajn, A. Bienick, and J. Z. Brzezinski, Bull. Acad. Pol. Sci., 1975, 23, 917).

2-(Trimethylsilylmethyl)pyridine is a useful synthon for the synthesis of a number of pyridines (T. Konakahara and Y. Takagi, Tetrahedron Letters, 1980, 2073). An improved preparation of the above silylpyridine via 2-pyridylmetyl-lithium has been reported (T. Konakahara and Y. Takagi, Heterocycles, 1980, 14, 393).

---

*For the chemistry of pyridylcarbenes see N. M. Lan and C. Wentrup, Helv., 1976, 59, 2068; C. Mayor and C. Wentrup, J. Amer. chem. Soc., 1975, 97, 7467 as well as W. D. Crow, A. N. Khan, and M. N. Paddon-Row, Austral. J. Chem., 1975, 28, 1741.

Interesting pyridylfurans can also be obtained *via* such sily-lated pyridines (O. Tsuge, K. Matsuda, and S. Kanemasa, Hete-rocycles, 1983, 20, 593). Diastereoselectivity of aldol-type reactions of alkylpyridines with benzaldehyde has been studied (H. Hamana and T. Sugasawa, Chem. Letters, 1984, 1591), it is possible to achieve 100% *erythro* selectivity.

(iii) Alkenylpyridines

Vinylpyridines can be used as versatile starting materials for cyclohexanones and hence steroids (S. Danishefsky, P. Chain, and A. Nagel, J. Amer. chem. soc., 1975, 97, 380) and (S. Danishefsky and P. Chain, J. org. Chem., 1975, 40, 3606). In this reaction sequence 3-vinylcyclohex-2-en-1-ones can be pre-pared starting from 2-vinylpyridines *via* reductive cyclisa-tion, for example

The 2-and 4-cyclopropylpyridines can be prepared, in 56-60% yield from the corresponding vinylpyridines and diazomethane (S. Al-Khaffaf and M. Shanshal, J. prakt. Chem., 1983, 325, 517).

(iv) Ethynylpyridines

2-Ethynylpyridine can be prepared from 2-methylbut-3-yn-2-ol (D. E. Ames, D. Bull, and C. Takundwa, Synthesis, 1981, 364).

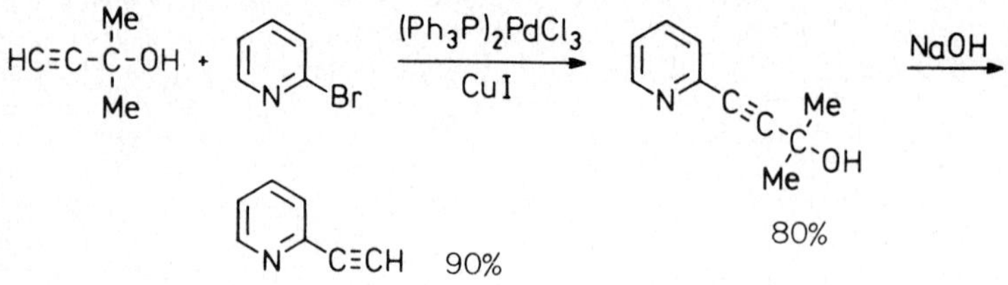

A similar reaction sequence can be used to produce 4–ethynyl-pyridine and subsequent oxidation of this compound yields 1,4-bis(4-pyridyl)butadiyne (L. D. Ciana and A. Haim, J. hetero-cyclic Chem., 1984, 21, 607).

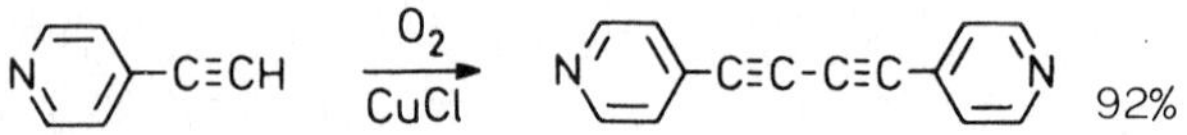

A variety of synthetic routes to pyridinylpropiolic esters has been reported (W. N. Kok and A. D. Ward., Austral. J. Chem., 1978, 31, 617).

(v) Arylpyridines

Chemical reactions of arylpyridines as well as heterocyclic substituted pyridines have been reviewed (H. Beschke, Aldrich-chimica Acta, 1981, 14, 13).

(vi) Pyridone methides (methylenepyridines)

2–Chloropyridinium salts react with malononitrile or ethyl cyanoacetate (R^1 = CN or CO_2Et) in triethylamine yielding the corresponding pyridone methides (H. Pauls and F. Kröhnke, Ber., 1977, 110, 1294).

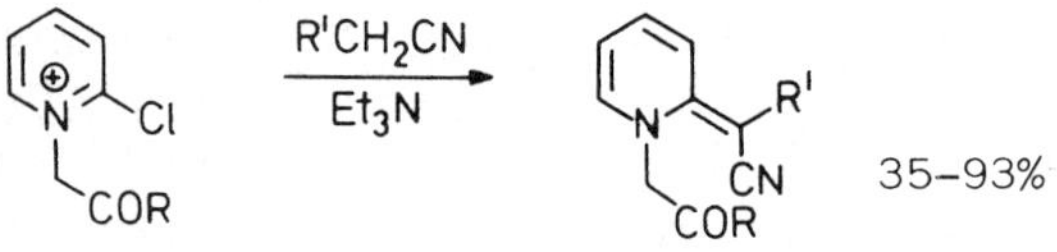

Such pyridine derivatives are useful for the preparation of indolizines (T. Uchida and K. Matsumoto, Synthesis, 1976, 209), a good example is the reaction shown below (P. Molina, P. M. Fresneda, and M. C. Lajara, J. heterocyclic Chem., 1985, 22, 113).

Thermally very stable 1-acyl or 1-sulphonyl-4-alkylidene-1,4-dihydropyridines are easily obtained from 4-benzylpyridines (E. Anders, W. Will, and A. Stankowiak, Ber., 1983, 116, 3192) and (E. Anders and A. Stankowiak, Synthesis, 1984, 1039).

The sulphonylderivatives are useful for the preparation of sulphonyl esters.

2-Allylidene-1,3-dihydropyridines are vinylogues of 2-methylene-1,2-dihydropyridines and can be prepared by condensation reactions of 2-methylpyridinium salts (A. Kakehi et al., J. org. Chem., 1978, 43, 4837 and references cited therein), for example in the preparation of indolizines.

(x) Reduction products of nitropyridines*

(1) Aminopyridines

Aminopyridines are an important type of pyridine derivatives as a large number of functionalizations can be carried out starting from such compounds.

A direct synthesis of N,N-dialkylaminopyridines from $2(1H)$- and $4(1H)$-pyridone by heating with a dialkylamine and phosphorus pentoxide has been described (E. B. Pedersen and D. Carlsen, Synthesis, 1980, 844). 4-Pyridyl-(aryl)amines can be prepared in excellent yields from 4-chloro-1-pyridiniopyridinium salts and primary or secondary amines (M. P. Sammes et al., J. chem. Soc. Perkin I, 1983, 973). A new method for imination of N-alkyl pyridinium salts by low temperature oxidation in liquid ammmonia with potassium permanganate has been reported (H. C. van der Plas and D. J. Buurman, Tetrahedron Letters, 1984, 3763).

R = alkyl   75–80%

The 4-iminoderivative can be obtained when R = Bu^t. The imines can be hydrolysed to the corresponding pyridones. The tetrafluoro-4-pyridyl-diazonium ion can be prepared by cautious diazotisation in acidic medium (A. C. Alty, R. E. Banks, and A. R. Thompson, J. fluorine Chem., 1984, 26, 263). Another type of reaction of 2- or 4-aminopyridine is the conversion into the corresponding formamide oximes with tris-formaminomethane followed by treatment with hydroxylamine (B. Stanovnik, I. Zmitch, and M. Tisler, Heterocycles, 1981, 16,

---

*Discussions on halogenopyridines, nitropyridines, and pyridine sulphonic acids (sections (vii, viii, and ix)) can be found in the sections dealing with electrophilic and nucleophilic reactions of pyridine in the present review. For long range $^{13}C-^1H$ coupling constants of aminopyridines, see Y. Takeuchi and N. Dennis, Org. mag. Res., 1975, 7, 244. For ^{15}N-NMR spectra of aminopyridines, see W. Städeli et al., Helv., 1980, 63, 504. For an MO study of 4-hydroxyaminopyridine 1-oxide tautomerism, see Y. Miyaji, H. Ichikawa, and M. Ogata, Chem. pharm. Bull., 1975, 23, 1256.

2173). Halogenation of 2-aminopyridine yields 2-amino-3,5-di-halogenopyridine, and controlled chlorination of 2,6-diamino-pyridine affords 2,6-diamino-3,5-dichloropyridine, useful for the preparation of 2,3,5,6-tetrachloropyridine (T. K. Chen and W. T. Flowers, Chem. Comm., 1980, 1139).

The corresponding bromination can also be carried out. A direct transformation of 3-aminopyridine to 3-styrylpyridine has been reported (K. Kikukawa et al., Chem. Letters, 1980, 551).

There are numerous examples of the use of especially pyridones and thiones as reagents in organic synthesis. Also aminopyridines can be synthetically useful*, for example unsymmetrical secondary alcohols can be prepared from Grignard reagents and 2-(N-methyl-N-formyl)aminopyridine (D. L. Comins and W. Dernell, Tetrahedron Letters, 1981, 1085).

For related reactions see (D. L. Comins and A. I. Meyers, Synthesis, 1978, 403; A. I. Meyers and D. L. Comjus, Tetrahedron Letters, 1978, 5179).

4-N,N-Dimethylaminopyridine is a more effective catalyst for the silylation of alcohols than pyridine itself (S. K. Chaudhary and O. Hermandez, Tetrahedron Letters, 1979, 99).

---

*For a review on 4-aminopyridines as catalyst in acylations, see G. Höfle, W. Steglich, and H. Vorbrüggen, Angew. Chem. intern. edn., 1978, 17, 569.

(2) Nitrosopyridines

A general method for the conversion of a primary amino group into a nitroso group for the synthesis of nitroso substituted heterocycles has been reported (E. C. Taylor, C. P. Tseng, and J. B. Rampal, J. org. Chem., 1982, 47, 552). Thus, 2-amino-pyridine and 2-amino-4-methylpyridine gave the corresponding nitrosopyridine by the following method (NCS = N-chloro-succinimide, MCPBA = m-chloroperbenzoic acid).

This extremely reactive nitroso compound condenses readily with 1,3-dienes and aromatic amines.

(3) Pyridyl isothiocyanates

An improved method for the synthesis of substituted 2-pyridyl isothiocyanates, yield 80-84%, starting from 2-aminopyridine, has been reported (D. J. Le Count, D. J. Drewsbury, and W. Grundy, Synthesis, 1977, 582).

(xi) Hydroxypyridines and pyridones*

Generally 2- and 4-pyridones, as well as 3-hydroxypyridine, are easy to prepare and as a consequence such pyridines are convenient starting materials for further functionalization into a range of substituted pyridines.

(1) Oxidation

Hexacyanoferrat(III) oxidation under basic conditions of 1-me-thyl-2-t-butylpyridinium salts results in elimination of the

---

*For various aspects of hydroxypyridine pyridone tautomerism, see P. Beak, Acc. chem. Res., 1977, 10, 186; M. L. Tosato, M. Cignitti, and L. Paoloni, Gazz., 1975, 105, 385; A. Maquestiau et al., Canad. J. Chem., 1975, 53, 490; R. S. Brown, A. Tse, and J. C. Vederas, J. Amer. chem. Soc., 1980, 102, 1174; M. J. Scanlan, I. H. Hillier, and R. H. Davies. Chem. Comm., 1982, 685.

t-butyl group and formation of 1-methyl-2-pyridone (H. Weber, Arch. Pharm., 1976, 309, 396; P. Nesvadba and J. Kuthan, Coll. Czech. chem. Comm., 1983, 48, 511). Oxidation of 1-substituted-2-methylpyridinium salts with pentyl nitrite under basic conditions yields the corresponding 2-pyridones (A. R. Katritzky and M. Shanta, Chem. Comm., 1979, 552). The smooth hydrolysis by aqueous sodium hydroxide of azabiphenylenes to pyridones reflects the increased reactivity of the pyridine rings in such systems (J. A. H. McBride and P. M. Wright, Tetrahedron Letters, 1982, 1109).

(2) Halogenation and halogeno pyridones

Nitryl chloride can be used for the chlorination in the 5-position of substituted 2-pyridones (M. V. Prostenik *et al.*, Croatica Chem. Acta, 1976, 48, 179). The bromination of 2-pyridone with aqueous bromine in the presence of KBr results in the formation of 3-bromo-2-pyridone (O. S. Tee and M. Paventi, J. Amer. chem. Soc., 1982, 104, 4142). 4-Pyridone yields pentabromopyridine after treatment with bromine in oleum and subsequent reaction with phosphorus oxybromide (S. D. Moshchitskii *et al.*, Chem. heterocyclic Comp. USSR, 1978, 55).

1-Fluoro-2-pyridone can be prepared by the reaction of 5% fluorine in nitrogen with 2-(trimethylsiloxy)pyridine in $FCCl_3$ at -78°C. 1-Fluoropyridine is a selective and safe fluorinating agent (S. T. Purrington and W. A. Jones, J. org. Chem., 1983, 48, 761).

4-Fluoro-2-pyridone can be prepared by ethercleavage of 4-fluoro-2-methoxypyridine (C. C. Leznoff *et al.*, J. heterocyclic Chem., 1985, 22, 145).

1-Hydroxy-2-pyridone can be prepared in 91% yield by oxidation of 2-(trimethylsilyloxy)pyridine with a Mo(V)peroxide-DMF complex (S. A. Matlin, P. G. Sammes, and R. M. Upton, J. chem.

Soc. Perkin I, 1979, 2481).

(3) Claisen rearrangements and related reactions

The ether from allyl bromide and 3-hydroxy-2-pyridone can
undergo a Claisen rearrangement (B. B. Jarvis and C. B.
Anderson, J. heterocyclic Chem., 1983, 20, 471).

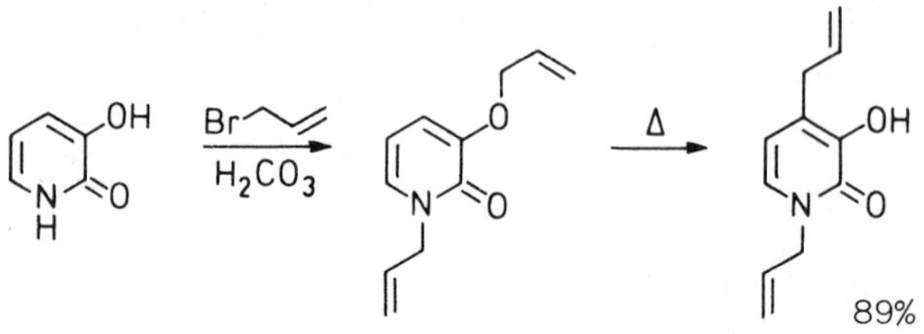

89%

2-Allyloxypyridine will rearrange in quantitative yield when
heated with a platinum catalyst, $Pt(PPh_3)_4$, in THF or DMF (G.
Balavoine and F. Guibe, Tetrahedron Letters, 1979, 3949).

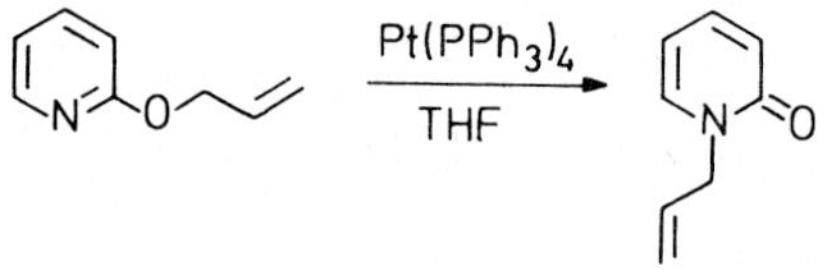

Various 1-benzyloxy or 1-allyloxy-2-pyridones rearrange thermally or photochemically yielding 3- or 3- and 5-disubstitu-
ted-2-pyridones with elimination of formaldehyde (A. R.
Katritzky *et al.*, J. chem. Soc. Perkin I, 1980, 2743), for
example

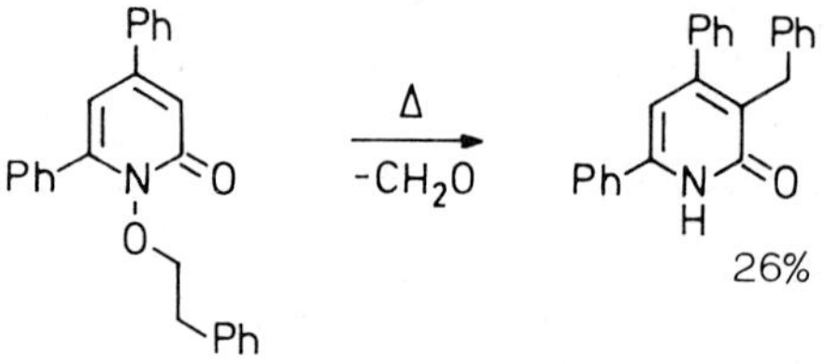

26%

(4) Alkylation and related reactions

Methyl or ethyl acrylate with palladium acetate can alkylate
2-pyridones (T. Itahara and F. Ouseto, Synthesis, 1984, 488).

34–56%

Functionalization of 2-pyridones can be performed by a novel dianion route (J. M. Domagala, J. heterocyclic Chem., 1984, 21, 1705).

A large number of other functionalizations of 2- and 4-pyridones can be found in the following references: J. M. Domagala, J. org. Chem., 1984, 49, 126; N. C. Hung and E. Bisagni, Synthesis, 1984, 765; T. Kametani et al., Chem. pharm. Bull., 1976, 24, 1870; A. A. Pessolano et al., J. heterocyclic Chem., 1985, 22, 265. They all demonstrate the versatility of the pyridones in organic synthesis.

Enantiomers of sterically hindered 1-aryl-4-pyridones have been separated and the barrier ($\Delta G^{\ddagger}$: 93–134 kJmol^{-1}) to partial rotation about the C–N bond determined (M. Mintas et al., Tetrahedron, 1985, 41, 229). 3,3'-oxybispyridine can be obtained in one step from 3-bromopyridine and 3-hydroxypyridine by heating with potassium carbonate and cuprous oxide at 190° (D. J. Barker and L. A. Summers, J. heterocyclic Chem., 1983, 20, 1411; this paper also gives a review for the other isomeric oxybispyridines).

(5) Pyridones as synthetic reagents

2-Pyridone is an excellent leaving group, therefore alkyl 2-pyridylcarbonates as well as 1-(alkoxycarbonyl)-2-pyridones are useful reagents for the introduction of protection groups into amino acids (F. Effenberger and W. Brodt, Ber., 1985, 118, 468; F. Effenberger, M. Keil, and E. Bessey, Ber., 1980, 113, 2110 and references cited therein).

$$R^1\text{-CH-CO}_2H \quad + \quad \text{(pyridine-OCO}_2R) \longrightarrow R^1\text{-CH-CO}_2H \ + \ \text{(2-pyridone)} \quad \sim 90\%$$

Related reactions are described by T. Mukaiyama, T. Masui, and T. Izawa (Chem. Letters, 1976, 1177; T. Keumi, R. Taniguchi, and H. Kitajuma, Synthesis, 1980, 139; for a review see T. Mukaiyama, Angew. Chem., 1979, 91, 798). 2-Pyridones may also be used for the preparation of aldehydes starting from alkyl halides (M. J. Cook, A. R. Katritzky, and G. H. Millet, Heterocycles, 1977, 7, 227).

Bis-2-Pyridylthionocarbonate prepared from thiophosgene and 2-pyridone is a useful reagent for the preparation of isothiocyanates and carbodiimides (S. Kim and K. Y. Yi, Tetrahedron Letters, 1985, 1661).

3,5-Dinitro-1-(p-nitrophenyl)-4-pyridone can be used as a protective group for primary amines and amino acids (E. Matsumura et al., Tetrahedron Letters, 1981, 757).

A mild and regioselective method for O-alkylation of the ambident 2-amino-3-hydroxypyridine by phase transfer catalysis has been reported (J. A. Bristol, I. Gross, and R. G. Lovey, Synthesis, 1981, 971). In an interesting reaction 3-hydroxypyridines can be carboxylated at high pressure to give pyridine carboxylic acids (F. Mutterer and C. D. Weis, J. heterocyclic Chem., 1976, 13, 1103).

(6) Cycloaddition reactions of pyridones*

2-Pyridones can react as dienes resulting in cycloaddition across the 3,6-positions (G. P. Gisby, S. E. Royall, and P. G. Sammes, J. chem. Soc. Perkin I, 1982, 169; K. Matsumoto *et al.*, Chem. Comm., 1979, 1091).

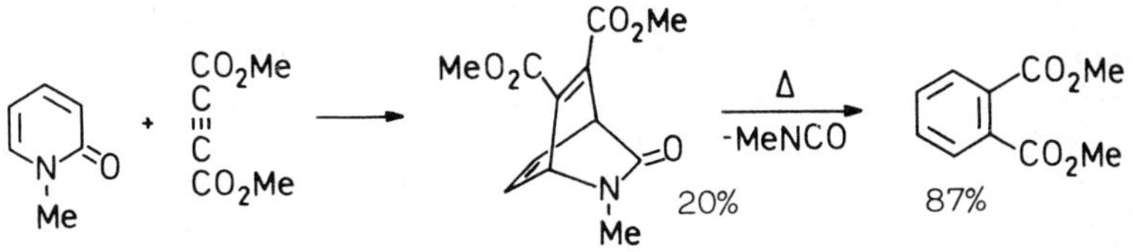

(xii) Pyridinethiols, pyridinethiones (thiopyridones), and their derivatives

Pyridinethiones can be prepared by a number of methods and are hence easily accessible compounds. It has been shown that derivatives of pyridinethiones are very useful and versatile reagents in organic synthesis. It has been demonstrated that the -SH tautomers of 2- and 4-pyridinethiones are the stable forms in gas-phase (C. B. Theissling *et al.*, Tetrahedron Letters, 1977, 1777).

(1) Preparation of pyridinethiones and related compounds

Preparation of pyridinethiones by ring synthesis has been de-scribed in the beginning of this chapter but pyridinethiones can also be obtained by various displacement reactions already described in the 2nd edition. New reactions include for example displacement of sulphoxide groups (N. Furikawa *et al.*, J. chem. Soc. Perkin I, 1984, 1839) and of a pyridinium cation by a sulphur nucleophile (B. Boduszek and J. S. Wieczorek, Monatsh., 1980, 111, 1111). A new method for the conversion of pyridinium-2-carboxylates into 2-pyridinethiones (A. R. Katritzky and H. M. Faid-Allah, Synthesis, 1983, 149) is shown below.

---

*The related photochemical reactions of pyridones can be found on page **58** .

35-72%

For the preparation of tetrachloro-2-pyridinethione, see B. Iddon *et al.*, J. chem. Soc. Perkin I, 1974, 2300).

2- and 3-Pyridyl sulphides are formed by deoxydative substitutions of pyridine 1-oxides by thiols in the presence of an acid chloride or anhydride (L. Bauer and S. Prachayasittikul, Heterocycles, 1986, 24, 161 and references cited therein). A stable pyridinesulphenyl halide has been reported (R. Matsueda and K. Aiba, Chem. Letters, 1978, 951).

100%

Pyridine *N*-sulphides have been isolated and unambiguously characterised for the first time (R. A. Abramovitch, A. L. Miller, and J. Pilski, J. chem. Soc. Perkin I, 1981, 703).

72%

New pyridinethiols and thiones have been prepared by various replacement reactions from pyridine derivatives.

(K. Krowichi, Pol. J. Chem., (a) 1978, 52, 2039; (b) 1979, 53, 503; (c) 1979, 53, 889). Compound a is unstable, beginning to evolve H_2S at 130°C and melting at 216-220°C d. b has m.p. 122-124°C and c has m.p. 210-212°C. 3,3'-Thiobispyridine can be obtained in 60% yield from 3-pyridinethiol and 3-bromo-pyridine (L. A. Summers and S. Trotman, J. heterocyclic Chem., 1984, 21, 917). A number of macrocyclic compounds possessing

2,6-pyridino sub-units connected by carbon-sulphur linkages have been prepared from 2,6-dihalopyridines (see for example G. R. Newkome *et al.*, J. org. Chem., 1978, 43, 2685).

Many new 3-(alkylthio)-2-halopyridines have been prepared by functionalization of various pyridine derivatives (G. S. Ponticello *et al.*, J. org. Chem., 1979, 44, 3080), oxidation of such alkylthiopyridines can produce sulphoxides or sulphones.

A number of other sulphoxide substituted pyridines have been prepared and used as phase-transfer catalysts (N. Furukawa *et al.*, J. chem. Soc. Perkin I, 1984, 1833).

(2) Cyclisation reactions from pyridinethiones

Starting from 2-pyridinethiones various cyclisation reactions have been reported, for example: thienopyridines (R. D. Youssefyeh *et al.*, J. med. Chem., 1984, 27, 1639; J. Becher *et al.*, Tetrahedron, 1978, 34, 989; P. Molina *et al.*, Synthesis, 1986, 342), thiazolopyridinium salts (K. A. M. Walker, E. B. Sjøgren, and T. R. Matthews, J. med. Chem., 1985, 28, 1673; P. Molina *et al.*, J. heterocyclic Chem., 1984, 21, 1609), thiadiazolopyridinium salts (P. Molina, A. Argues, and H. Hernandez, Synthesis, 1983, 1021).

(3) Reactions of pyridinethiones and related compounds

Pyridinethiones have been used extensively as reagents for various acylation and alkylation reactions. Such reactions are based on the stability of the heteroaromatic 2-pyridinethione system which renders it an excellent leaving group. This is the reason for the use of 2-pyridyl sulphides (sepharose-glu-tathione-2-pyridyl disulphide) in affinity chromatography of thiol containing peptides ("Affinity Chromatography, Prin-ciples and Methods", page 36, Pharmacia Fine Chemicals AB, Uppsala, Sweden), the scheme below outlines the principle involved.

Usually 2-pyridinethiol carboxylic esters are prepared from 2,2'-dipyridyl sulphide. A new method is the use of 2-thio-pyridyl chloroformate (E. J. Corey and D. A. Clark, Tetrahedron Letters, 1979, 2875).

Such 2-pyridinethiol carboxylic esters are becoming increasingly important as acylating agents in the synthesis of peptides, ketones, macrocyclic lactones, and β-lactams (see E. J. Corey and D. A. Clark, *loc. cit.* ; S. Kobayashi *et al.*, J. Amer. chem. Soc., 1981, 103, 2406; S. Kim, J. I. Lee, and B. Y. Chung, Chem. Comm., 1981, 1231).

In the presence of oxygen a carboxylic ester is formed, while the reaction under nitrogen yields a ketone. Another example is the acylation of pyrroles for natural product synthesis (K. C. Nicolaou, D. A. Claremon, and D. P. Papahatjis, Tetrahedron Letters, 1981, 4647).

R = alkyl or aryl

For related C-glycosidation of pyridyl thioglycosides see R. M. Williams and A. O. Stewart, Tetrahedron Letters, 1983, 2715 and references cited therein.

An important method for initiating radical chain reaction by irradiation of 1-acyloxy-2-pyridinthiones has been developed (D. H. R. Barton, H. Togo, and S. Z. Zard, Tetrahedron, 41, 1985, 5507 and references cited therein). The method avoids the usual complication due to polymerisation in radical reactions, the overall reaction follows the mechanism given below.

The radical formed can be intercepted by a variety of reagents, for example nitro-olefins (a). The nitrosulphides (Z = NO_2) (b) thus produced subsequently yield ketones by reductive cleavage by $TiCl_2$. Dialkylaminyl radicals can also be generated via 1-hydroxypyridine-2-thione carbamates (M. Newcomb et al., Tetrahedron Letters, 1985, 5651).

(4) Nucleophilic substitution of alkylthio groups

The alkylthio group in 2-alkylthiopyridinium salts can be displaced by nucleophiles, for example in the preparation of thiols (M. Yamada et al., J. org. Chem., 1977, 42, 2180) or 2-methylenepyridines (P. Molina and A. Lorenzo, Tetrahedron Letters, 1983, 5805). A 4-methylthio group in 4-methylthiopyridines can be replaced by hydrogen, alkyl, or aryl groups via Grignard reactions (E. Wenkert et al., J. org. Chem., 1985, 50, 1125).

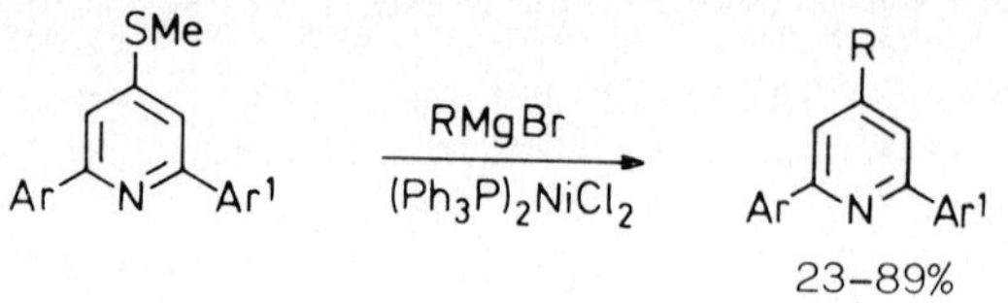

84

23–89%

(5) Rearrangement reactions of pyridinethiones and related compounds

2-Tetrazolylthio-3-aminopyridines can undergo a Smiles rearrangement (H. W. Altland, J. org. Chem., 1976, 41, 3395).

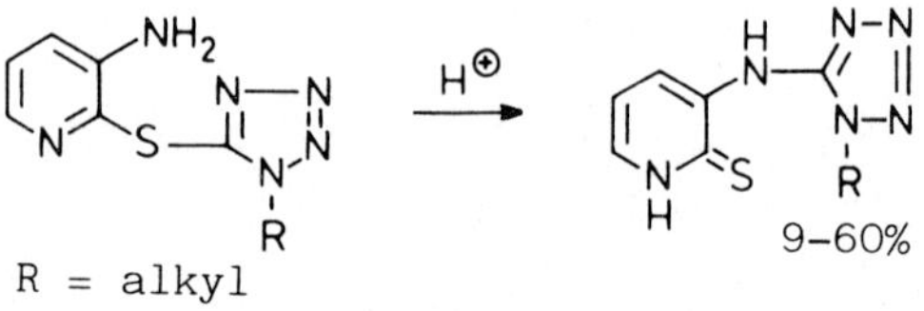

R = alkyl

9–60%

Pyridyl sulphides can be prepared by Chapman rearrangement of 2-pyridinethione salts (P. Molina *et al.*, Synthesis, 1982, 598).

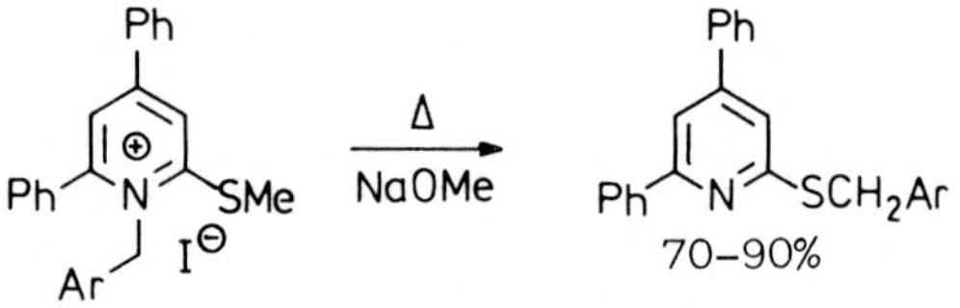

70–90%

(6) Pyridineselones and related compounds*

Synthesis of 1-substituted-3-formyl-2-pyridineselones has been reported (J. Becher *et al.*, Acta Chem. Scand., 1977, B31, 843). 2-Pyridylseleno carbonyl compounds can be used for enone synthesis as described above in section *(4)* for the corresponding thiones (A. Toshimitsu *et al.*, Tetrahedron Letters, 1982, 2105). The preparation of 1-hydroxy-2-pyridineselones and their use as fungicides and bactericides has been described (R. Henderson, F. Rothgery, and H. A. Schroeder, U. S. Patent, 1985, 4.496,559; Chem. Abs., 1985, 102, 166621c).

---

*For a review on selenium pyridine compounds, see H. L. Yale, Chem. heterocyclic Comp., 1975, 14, Part 4, 189.

(xiii-xv) Pyridyl alcohols, pyridine carbaldehydes. and pyridyl ketones

New and useful ring syntheses for pyridine-3-carbaldehyde derivatives as well as for various pyridyl ketones can be found in the first part of this chapter. Ring syntheses of 3-acyl or 3-formyl pyridine derivatives are relatively important as direct introduction, for example by a Friedel-Crafts reaction, is not possible. However, the preparation of 2-alkyl-3-acyl or 3-formylpyridines *via* a Sommelet-Hauser rearrangement of α-pyrrolidinyl-2-alkylpyridine has been reported (E. B. Sanders, H. V. Secor, and J. I. Seeman, J. org. Chem., 1978, 43, 324).

A more direct method of introducing a formyl group is the hydrolysis of a dibromomethyl group (K. Eichinger, H. Berbalk, and H. Kronberger, Synthesis, 1982, 1094).

Formyl and acylpyridines are useful reagents* in organic synthesis, for example for the preparation of nicotine analogues and other pyridine alkaloides, (see for example T. E. Catha

---

*For reactions of various pyridine aldehydes, see H. Stetter, Angew. Chem. intern. ed., 1976, 15, 639.

and E. Leete, J. org. Chem., 1978, <u>43</u>, 2125; F. Benington, R. D. Morin, and M. A. Khaled, Synthesis, 1984, 619; J. Becher, T. Johansen, and M. A. Michael, J. heterocyclic Chem., 1984, <u>21</u>, 41).

(xvi) Pyridine carboxylic acids

A high yield decarbalkoxylation of various alkyl 2- or 6-alkoxynicotinoates has been reported (G. R. Newkome, D. K. Kohli, and T. Kawato, J. org. Chem., 1980, <u>45</u>, 4508), for example.

$$280°C \qquad 100\%$$

2-Alkoxynicotinamides undergo rearrangement without loss of the amide group.

$$280°C \qquad 100\%$$

Mercuration of pyridine-2,3-dicarboxylic acid results in formation of a 2-mercurated pyridine (T. Takahashi, Chem. pharm. Bull., 1979, <u>27</u>, 2473).

$$i) NaOH$$
$$ii) HgO/AcOH$$

All four isomeric pyridine analogues of phthalide have been prepared from approprate pyridine dicarboxylic acids, making use of the differential reactivity of the pyridine substituents (W. R. Ashcroft, M. G. Beal, and J. A. Joule, J. chem. Soc. Perkin I, 1981, 3012).

Pyridine 2-, 3- or 4-carbohydroxamic acids may be converted into the corresponding aminopyridine carboxylic acids *via* a modified Lossen rearrangement (Z. Eckstein, E. L. Kochany, and J. Krzeminski, Heterocycles, 1983, <u>20</u>, 1899).

$$\Delta$$
$$HCONH_2 \qquad 84\%$$

Irradiation of 2-azidocarbonylpyridinium salts can be used for preparation of aryl aldehydes (A. R. Katritzky and T. Siddiqui, J. chem. Soc. Perkin I, 1982, 2953) *via* rearrangement of a nitrene intermediate.

Treatment of 3-hydroxy-6-methylpyridine-2-carboxylic acid with ammonia yields the corresponding carboxamide and not the 3-aminopyridine as previously claimed (G. G. I. Moore, A. R. Kirk, and R. A. Newmark, J. heterocyclic Chem., 1979, 16, 789).

Interestingly, it is possible to separate the two enantiomerically pure 2,4-dimethyl-N,N-dimethyl-3-carbamoylpyridines, as steric hindrance prevents racemisation (P. M. van Lier, G. H. W. M. Meulendijks, and H. M. Buch, Rec. Trav. Chim., 1983, 102, 337).

(xvii) Cyanopyridines

Cyanopyridines are important compounds for the preparation of biologically active nicotinamide and isonicotinamide derivatives. 3-Cyanopyridine is usually prepared industrially by ammonoxidation of 3-methylpyridine. A review of the patent literature has been published (H. Beschke *et al.*, Chemiker-Zeitung, 1977, 101, 384).

A laboratory ammonoxidation of 4-methylpyridine giving 4-cyanopyridine in 98% yield has been reported (C. H. Wang, F. V. Hwang, and J. M. Horny, Heterocycles, 1979, 12, 1191). An efficient synthesis of 4-cyanopyridine *via* the easily prepared 1-(arylazodiphenylmethyl)pyridinium bromide has been reported (J. Schantl and H. Gstach, Synthesis, 1980, 694).

Both the 2- and 4-cyanopyridines can be reduced to pyridine on treatment with aqueous titanium trichloride (A. Clerici and O. Porta, Tetrahedron, 1982, 38, 1293), for 4-cyanopyridine the yield is quantitative. Pyridine carboxamides are often used for the preparation of cyanopyridines, (L. I. M. Spiessens and M. J. O. Ateunis, Bull. Soc. chim. Belg., 1980, 89, 205). Many cyanopyridines may also be obtained directly *via* ring synthesis. Consequently, cyanopyridines are frequently used for conversion into other pyridine derivatives, such as imino ethers (J. L. La Mattina and R. L. Taylor, J. org. Chem., 1981, 46, 4179).

The very unstable pyridine-3-nitrile oxide has been prepared from pyridine-3-aldoxime chloride and triethylamine (M. Majeurski, B. Serafin, and T. Oklesinska, Roczniki Chimii, 1977, 51, 975).

5. Bi-, ter-, quater-pyridyls, other polypyridyls, and macrocyclic systems containing pyridines (pyridinophanes)*

The di-quaternary salts of bipyridines are well investigated due to their herbicidal activity, furthermore there is much interest in the study of multidentate and selective ligands incorporating pyridines into a macrocyclic framework.

For a discussion on the preparation of 2,2'-bipyridyls see K. T. Potts and P. A. Winslow, J. org. Chem., 1985, 50, 5405. This paper also describes a new method for the preparation of 2,2'-bipyridyls *via* α -oxoketene dithioacetals. Examples of optically active 2,2'-bipyridyls have been described (C. Botteghi, G. C. Chelucci, and F. Soccolini, J. org. Chem., 1984, 49, 4290). Various 2,2'-, 3,3'-, and 4,4'-bipyridines can also be prepared by nickel complex mediated coupling reactions of halopyridines (M. Tiecco *et al.*, Synthesis, 1982, 736). For the use of 2-pyridyl sulphoxides in the synthesis of 2,2'-bipyridines see T. Kawai, N. Furikawa, and S. Oae, Tetrahedron Letters, 1984, 2549. An improved method for the

---

*Comprehensive reviews on the chemistry of polypyridyls have been published.

V. K. Majestic and G. R. Newkome, Topics in Current Chem., 1982, 106, 79, "Pyridinophanes, Pyridinocrowns, and Pyridinocryptands"; G. R. Newkome, V. H. Gupta, and J. D. Sauer, The Chem. of Heterocyclic Comp., Pyridine and its Derivatives, part 5, G. R. Newkome ed., J. Wiley, N. Y., 1984, 447; L. A. Summers, "The Bipyridinium Herbicides", Academic Press, N. Y., 1980; "The Electrochemistry of the Viologens", C. L. Bird and A. T. Kuhn, Chem. soc. Rev., 1981, 10, 49; "Macrocyclic pyridines", G. R. Newkome *et al.*, Chem. Rev., 1977, 77, 513; "Biter--, and oligopyridines", F. Kröhnke, Synthesis, 1976, 1.

preparation of the ter-pyridine *a* starting from 2-acetylpyri-
dine has been reported (K. T. Potts *et al.*, Org. Synth., 1986,
64, 189).

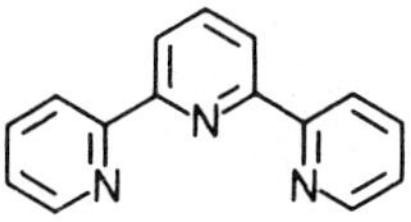

a

For other examples of ter-pyridine preparations see W. Spahni
and G. Calzaferri, Helv., 1984, 67, 450. The interesting
so-called "sexipyridine" *b* has been prepared by a useful
methodology starting from 6,6'-dimethyl-2,2'-bipyridine (G. R.
Newkome and H. W. Lee, J. Amer. chem. Soc., 1983, 105, 5956).

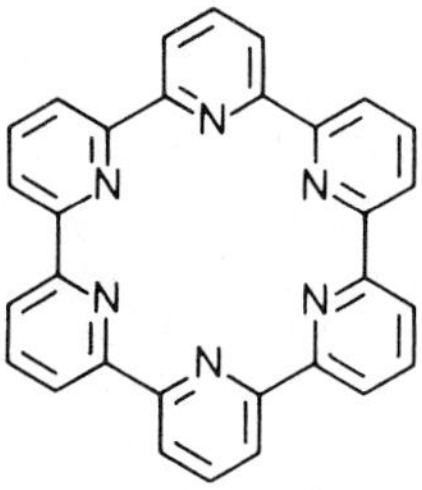

b

A large number of pyridine macrocyclic compounds have been
prepared (reviews *loc. cit.*). A typical procedure is illustrated by the following example (M. P. Coke, J. org. Chem.,
1981, 46, 1747).

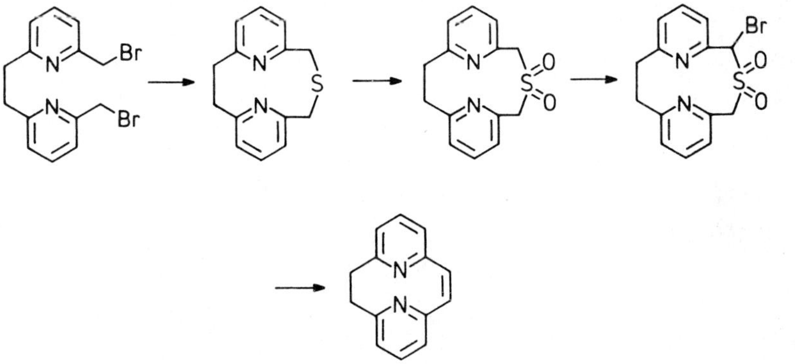

For the incorporation of bi- and ter-pyridines into various
diimine ligands and crown-ethers see E. C. Constable and J.

Lewis, Polyhedron, 1982, 1, 303; E. Buhleir, W. Wehner, and F. Vögtle, Ber., 1978, 111, 200; Ann., 1978, 537).

6. Dehydropyridines

(a) o-Didehydropyridines (pyridynes)

(i) Formation

As mentioned in the 2nd edition, the most stable of the pyridynes is the 3,4-pyridyne.

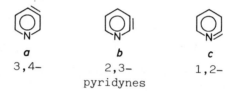

a	b	c
3,4-	2,3-	1,2-

pyridynes

The 2,3-pyridyne b is less stable and consequently this isomer has been less investigated. The relative stability of 3,4-pyridyne has been confirmed by MNDO calculations which furthermore suggest that biradical forms of 2,4-, 2,5-, and 2,6-didehydropyridine are comparable in stability with the known 2,3- and 3,4-pyridynes (M. J. S. Dewar and G. P. Ford, Chem. Comm., 1977, 539). Existence of the 1,2-pyridyne c has been questioned. It may be better represented by a pyridyl cation structure d (J. F. Bunnett and P. Singh, J. org. Chem., 1981, 46, 4567).

In fact decomposition of the zwitterionic precursor in the presence of various dienes does not give any cycloaddition products but only the normal products expected from electrophilic substitution reactions.

Displacement of a halogen atom from the 2- or the 4-position in pyridine by an amide ion proceeds *via* elimination-addition (a pyridyne EA-mechanism) or *via* addition-elimination (an AE-mechanism). In pyridine *N*-oxides a 2-halogen reacts *via* a pyridyne (EA-mechanism) while a 4-halogen reacts *via* an addition-elimination (AE-mechanism). In o-dihalogenopyridines lithiation leads to pyridynes. It is possible to o-lithiate 2-, 3-, and 4-halopyridines with lithium-diisopropylamide or butyllithium, thus 3,4-pyridyne can be generated by the reaction of 3-chloro-4-iodopyridine with butyllithium and trapped with furan (38% yield of Diels-Alder product) (G. W. Gribble and M. G. Saulnier, Tetrahedron Letters, 1980, 4137). Critical reviews on heteroaryne chemistry have appeared (M. G. Reinecke, Tetrahedron, 1982, 38, 427; H. C. van der Plas and F. Roeterdink, "The Chemistry of Triple Bonded Groups, Part 1, eds. S. Patai and Z. Rappoport, Wiley, New York, 1983, 421). The problem is that many heteroarynes are often difficult to detect as a typical *cine*-substitution product can arise *via* a number of possible mechanistic pathways without involving a heteroaryne. Even cycloaddition reaction products of hetero-arynes can sometimes be explained without the participation of a heteroaryne.

An improved and convenient method for the preparation of 3,4-pyridyne has been reported (C. May and C. J. Moody, Tetra-hedron Letters, 1985, 2123). The stable precursor is prepared from 3-aminopyridine-4-carboxylic acid.

(ii) Reaction of pyridynes

An important and novel application of a 3,4-pyridyne, prepared from 1-aminotriazolo[4,5-c]pyridine, is the use of this pyri-dyne in the construction of derivatives of the alkaloid ellip-ticine, the pyridine ring is introduced in one step (G. W. Gribble *et al.*, J. org. Chem., 1984, 49, 4518).

An even shorter convergent synthesis of ellipticine and isoel-lipticine (1:1, 40% total yield) is the following elegant method (C. May and C. J. Moody, Chem. Comm., 1984, 926).

Intramolecular annelation reactions involving pyridynes do not appear to have been much studied, although some interesting examples have been reported (S. V. Kessar *et al.*, Tetrahedron Letters, 1976, 3207; S. V. Kessar, Acc. chem. Res., 1978, 11, 283 and references cited therein).

This and related reactions need further study in order to prove the intermediacy of a pyridyne intermediate (H. C. van der Plas and F. Roeterdink, *loc. cit.*).

7. Hydrogenated pyridine derivatives*

(a) Dihydropyridines**

(i) Preparation

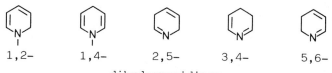

| 1,2- | 1,4- | 2,5- | 3,4- | 5,6- |

dihydropyridines

Usually 1,2- and 1,4-dihydropyridines can be obtained by reductive methods (E. Booker and K. Eisner, J. chem. Soc. Perkin I, 1975, 929; M. Acheson and G. Paglietti, *ibid.*, 1976, 45; D. L. Comins and A. H. Abdullah, J. org. Chem., 1984, $\underline{49}$, 3392), the last reference reports a regiospecific addition of hydride ion to the 4-position.

$$\text{Li(Bu}^t\text{O)}_3\text{AlH/CuBr} \qquad 20\text{-}65\%$$

R = H, alkyl, Cl, CO_2Me

In the presence of amines photochemical reduction of pyridines can give 1,2-dihydropyridine dimers (K. Kano and T. Matsuo, Tetrahedron Letters, 1975, 1389). In an alternative to the reductive methods, the parent 1,2-dihydropyridine can be prepared from its stable valence isomer (J. N. Bonfiglio *et al.*, J. Amer. chem. Soc., 1976, $\underline{98}$, 2344; P. Beeken *et al.*, *ibid.*, 1979, $\underline{101}$, 6677).

---

*Preparation of various reduced pyridines by metal hydrides has been reviewed (J. G. Keay, Adv. heterocyclic Chem., 1986, $\underline{39}$, 2).
**For reviews on dihydropyridines see D. M. Stout and A. I. Meyers, Chem. Rev., 1982, $\underline{82}$, 223; J. P. Kutney, Heterocycles, 1977, $\underline{7}$, 593; J. Kuthan and A. Kurfurst, Ind. Eng. Chem. Prod. Res. Dev., 1982, $\underline{21}$, 191. For various theoretical calculations on the stability of dihydropyridines, see N. Bodor and R. Pearlman, J. Amer. chem. Soc., 1978, $\underline{100}$, 4946; S. Böhm and J. Kuthan, Coll. Czech. chem. Comm., 1981, $\underline{46}$, 2068.

1,2-Dihydropyridines can be rearranged into the corresponding more stable 1,4-isomers by strong base or by transition metal catalysis (K. Eisner and M. M. Sadeghi, Tetrahedron Letters, 1978, 299 and references cited therein).

As already mentioned in the section on nucleophilic reactions of pyridinium compounds (page **48**), metallorganic reagents, such as Grignard-reagents, yield 1,2- and 1,4-dihydropyridines with activated pyridinium compounds. For useful examples see D. L. Comins and A. H. Abdullah, J. org. Chem., 1982, **47**, 4315; D. L. Comins, Tetrahedron Letters, 1983, 2807.

The regioselectivity in this type of reaction is very dependent upon R and R^I, with copper(I) iodide catalysis exclusive 1,4-addition takes place.

A 94% regioselective 1,2-allylation has been reported, yielding starting materials for alkaloid syntheses (R. Yamaguchi et al., J. org. Chem., 1985, **50**, 287 and references cited therein).

For other examples for the preparation of dihydropyridines *photochemically* see L. F. Tietze, A. Bergmann, and K. Bruggemann, Tetrahedron Letters, 1983, 3579 and references cited therein. For preparations *via* the *Hantzsch synthesis* see F. Bossert, H. Meyer, and E. Wehinger, Angew. Chem., intern. Edn., 1981, 20, 762. Various preparations *via* *nucleophilic reactions* on pyridines can be found in papers by A. E. Hauck and C. S. Giam, J. chem. Soc. Perkin I, 1980, 2070; M. G. El Din, E. E. Knaus, and C. S. Giam, Can. J. Chem., 1982, 60, 1821; A. I. Meyers and R. A. Gabel, J. org. Chem., 1982, 47, 2633; A. I. Meyers and N. R. Natale, Heterocycles, 1982, 18, 13, while various examples of routes from pyridinium compounds are described by K. Akiba, Y. Iseki, and M. Wada, Bull. chem. Soc. Japan, 1984, 47, 1994; K. Akiba, Y. Nishihara, and M. Wada, Tetrahedron Letters, 1983, 24, 5269; K. Akiba, Y. Iseki, and M. Wada, *ibid.*, 1982, 23, 429; R. Yamaguchi, Y. Nakazono, and M. Kawanisi, *ibid.*, 1983, 24, 1801; D. L. Comins and A. H. Abdullah, J. org. Chem., 1982, 47, 4315; M. J. Wanner, G. J. Koomen, and U. K. Pandit, Tetrahedron, 1982, 38, 2741; K. Akiba, Y. Iseki, and M. Wada, Tetrahedron Letters, 1982, 23, 3935.

An improved synthesis of a pyridine Reissert-analogue from pyridine trimethylsilyl cyanide and ethyl chloroformate has been reported (J. Kand and F. D. Popp, Chem. and Ind., 1985, 125).

For R = 3-Br the 4-isomer is formed, while other substituents (3-CN, 3-Me, 4-Me, 2-CH_2Ph, and 3,4,5,-Cl_3) result in formation of the 2-isomer.

The parent 2,5-, 3,4-, and 5,6-dihydropyridines are less well investigated due to their lower stability as compared to that of the 1,4-isomers. However, a 2,5-dihydropyridine can be obtained from 2-fluoropyridine with a butyllithium complex (F. Marsais, P. Granger, and G. Quequiner, J. org. Chem., 1981,

46, 4494).

65%

On distillation 2,5-dihydropyridines will disproportionate (R. F. Francis, C. D. Crews, and B. S. Scott, J. org. Chem., 1978, 43, 3227).

2,5-Dihydropyridines are also useful in the syntheses of alkaloids (E. Leete and M. E. Mueller, J. Amer chem. Soc., 1982, 104, 6440). The formation of a 3,4-dihydropyridine has been reported (J. C. Day, J. org. Chem., 1978, 43, 3646). Lithium metal reduction of 2,6-di-*t*-butylpyridine yields (96%) a mixture of 2,6-di-*t*-butyl-1,4- and 2,6-di-*t*-butyl-3,4-dihydropyridine.

The parent 5,6-dihydropyridine has been prepared by dehydrochlorination of the *N*-chloro precursor (M. C. Lasne *et al.*, Tetrahedron Letters, 1984, 3847 and references cited therein), NCS = *N*-chlorosuccinimide.

The parent 5,6-dihydropyridine has also been prepared by thermolysis of a bicyclic precursor.

5,6-Dihydropyridinium salts can be prepared from pyridinium salts by borohydride reduction followed by oxidation *via* the Polonovski reaction (D. S. Grierson, M. Harris, and H. P. Husson, J. Amer. chem. Soc., 1980, 102, 1064).

Such salts are useful synthons in alkaloid syntheses (M. Bonin *et al.*, J. org. Chem., 1984, 49, 2392 and references cited therein).

(ii) Properties and reactions

As already mentioned above, dihydropyridines are very important as starting materials for alkaloid syntheses*. An illustrative example among many uses of a 1,2-dihydropyridine is
the following (P. Beeken *et al.*, J. Amer. chem. Soc., 1979,
<u>101</u>, 6677).

For related examples of such annelations leading to complex
heterocyclic systems see B. Weinstein, L. C. C. Lin, and F. W.
Fowler, J. org. Chem., 1980, <u>45</u>, 1657; I. Hasan and F. W.
Fowler, J. Amer. chem. Soc., 1978, <u>100</u>, 6696; R. M. Wilson, R.
A. Farr, and D. J. Burlett, J. org. Chem., 1981, <u>46</u>, 3293; M.
J. Wanner, G. J. Koomen, and U. K. Pandit, Tetrahedron, 1983,
<u>39</u>, 3673).
  The following example is an intramolecular 4-addition,
yielding a fragment useful in morphine synthesis (D. D.
Weller, G. R. Luellen, and D. L. Weller, J. org. Chem., 1983,
<u>48</u>, 3061).

The investigation of the mechanism of NADH reductions as well
as the use of dihydropyridines that can mimic such reductions
are active areas of research, for a review see R. J. Hill and
D. A. Widdowson, "Bioorganic Chemistry", E. E. van Tamelen
ed., Academic Press, N. Y., 1978, <u>4</u>, 239. An interesting example is the enantiospecific reduction with a chiral 1,4-dihy-
dropyridine (A. Ohno *et al.*, J. Amer. chem. Soc., 1979, <u>101</u>,
7036).

---

*For a review on pyridine alkaloids, see C H. Eugster,
Heterocycles, 1976, <u>4</u>, 51.

94–97% optical yield

See also A. I. Meyers *et al.*, Tetrahedron Letters, 1981, 5123 and references cited therein as well as in B. J. van Keulen and R. M. Kellogg, J. Amer. chem. Soc., 1984, 106, 6029. A polymer bound NADH model has been prepared and used for reduction of benzaldehydes (R. Mathis *et al.*, Tetrahedron Letters, 1981, 59). Other transfer reactions can also be performed *via* 1,4-dihydropyridines, such as thio-substituted-1,4-dihydropyridines which can be used as thiolate transfer reagents (O. Piepers and R. M. Kellogg, Chem. Comm., 1980, 1147).

R = alkyl, aryl

a

b

The depicted macrocyclic bislactone crown ether *b* is an example of an enolate transfer reagent (see S. H. Mashraqui and R. M. Kellogg, J. Amer. chem. Soc., 1983, 105, 7792 and references cited therein).

1-Benzyl-1,4-dihydronicotinamide can be used as a reagent for replacement of aliphatic nitro groups by hydrogen (N. Ono, R. Tamura, and A. Kaji, J. Amer. chem. Soc., 1980, 102, 2851).

An interesting 3-oxopyridine is formed in the photo oxygenation of 2,4,4,6-tetraphenyl-1,4-dihydropyridine (K. Maeda, M. Nakamura, and M. Sakai, J. chem. Soc. Perkin I, 1983, 837).

49%

(b) Tetrahydropyridines (piperideines)*

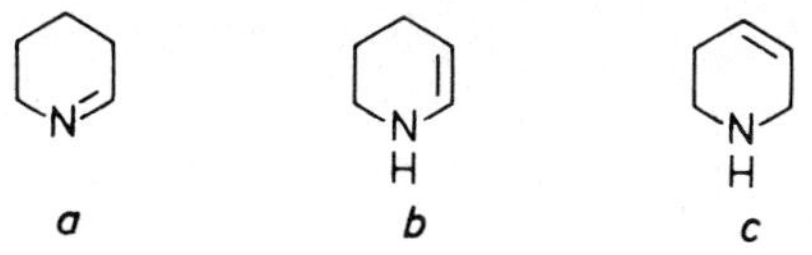

a	*b*	*c*
2,3,4,5-	1,2,3,4-	1,2,3,6-

tetrahydropyridines

(i) Preparation

The parent 1-piperideine *a* can be prepared from piperidine and is useful for example for the synthesis of (+)-anabasine (F. E. Scully, J. org. Chem., 1980, 45, 1515).

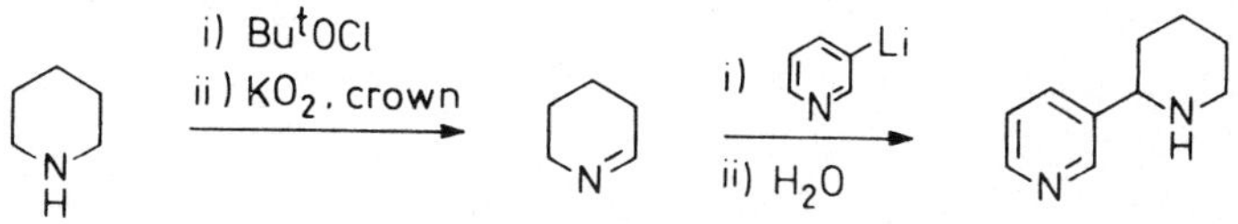

2-Piperideines can be obtained from 1,4-dihydropyridines by reduction. A convenient method using triethyl silane reduction is compatible with a range of substituents in the pyridine ring (U. Rosentreter, Synthesis, 1985, 210).

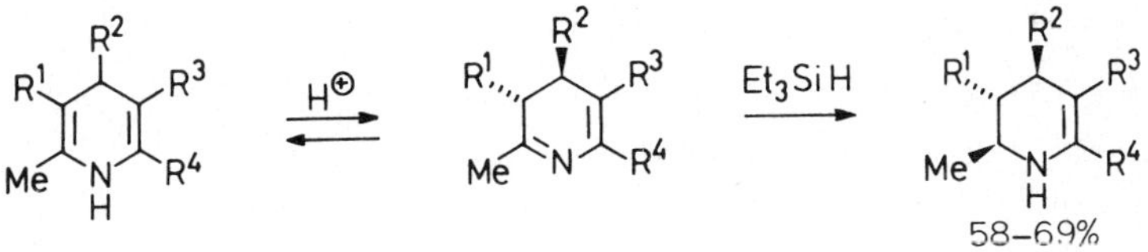

58–69%

The enamine *N*-methyl-2-piperideine can be prepared from the easily accessible *N*-methyl-1-piperideine by treatment with strong base (P. Beeken and F. W. Fowler, J. org. Chem., 1980, 45, 1336).

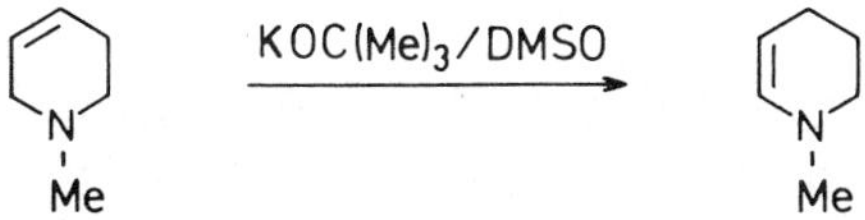

---

*Methyl vinyl ketone undergoes annelation reactions with 2-pi-perideines, a reaction type useful in alkaloid syntheses; R. V. Stevens, Acc. chem. Res., 1977, 10, 193; see also R. V. Stevens and N. Hrib, Chem. Comm., 1983, 1422.

2-Alkoxy-1-piperideines are easily prepared from 2-piperidones, the following examples demonstrate the method (M. M. Gugelchuk, D. J. Hart, and Y. M. Tsai, J. org. Chem., 1981, 46, 3671; J. P. Celerier et al., J. org. Chem., 1979, 44, 3089).

For the preparation of 5-benzylidene-1-piperideine from the 2-piperideine trimer see Y. Nomura et al., Tetrahedron Letters, 1979, 3453. Such 5-benzylidene-1-piperideines react with nucleophiles yielding 1,2-adducts (Y. Nomura et al., Bull. chem. Soc., 1984, 57, 1271).

N-Methylpipecolic acid gives the N-methyl-1-piperideinium salt with POCl₃ (R. T. Dean, H. C. Padgett, and H. Rappoport, J. Amer. chem. Soc., 1976, 98, 7448).

The corresponding N-oxides (nitrones) are also potential synthons for natural product syntheses (E. Gössinger, Monatsh., 1982, 113, 339 and references cited therein).

N-Aminopyridinium ylides yield the corresponding N-amino-3-piperideines with borohydride (E. E. Knaus and K. Redda, J.

heterocyclic Chem., 1976, 13, 1237).

R = PhSO$_2$-, PhCO-, MeCO-

Borohydride reduction of 1-substituted-3-oxidopyridinium betaines yields the 3-hydroxypiperidines while lithium aluminium
hydride yields 5-hydroxy-3-piperideines albeit in low yields
(W. R. Ashcroft and J. A. Joule, Heterocycles, 1981, 16,
1883).

R = aryl

Another potential synthon for compounds related to natural
products is the 5-oxo-3-piperideine system, prepared in a
7-step synthesis starting from 3-piperideine (T. Imanishi *et
al.*, Tetrahedron Letters, 1981, 4001; T. Imanishi *et al.*,
Chem. pharm. Bull., 1982, 30, 3617).

R = CO$_2$Et, SO$_2$Me

For a shorter route to this system see L. L. Chen *et al.*,
Heterocycles, 1984, 22, 2769. This reference also gives access
to related work in this area.

A simple large scale preparation of the related *N*-benzoyl-
4-oxo-2-piperideine has been reported (F. M. Schell and P. R.
Williams, Synthetic Comm., 1982, 12, 755). A variety of
3,4-dihydropyridin-2-ones can be prepared by the following
method (B. Sain, J. N. Baruah, and J. S. Sandhu, J. heterocyclic Chem., 1982, 19, 1511).

R = benzyl, aryl

68-90%

(ii) Properties and reactions

The 1-piperideine is an imine while the 2-piperideine contains an enamine function. The reactivity of such groups has been used in numerous syntheses of compounds related to natural products, of which some examples have already been mentioned in the previous section.

The following reaction sequence for the preparation of the alkaloid vincamone is typical (G. Massiot, F. S. Oliveira, and J. Levy, Tetrahedron Letters, 1982, 177).

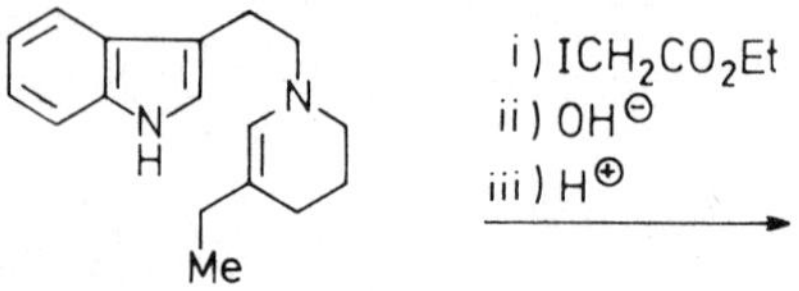

i) ICH_2CO_2Et
ii) $OH^\ominus$
iii) $H^\oplus$

For intramolecular Diels-Alder reactions of 2-piperideines see M. E. Kuehne *et al.*, J. org. Chem., 1980, 45, 3259; S. F. Martin *et al.*, J. Amer. chem. Soc., 1980, 102, 3294.

Δ

48%

For an intermolecular Diels-Alder reaction of the parent 1-piperideine see S. M. Weinreb and J. L. Levin, Heterocycles, 1979, 949.

The 3-piperideine system is the most stable of the piperideines. These compounds, which are easily prepared, are important starting materials for the preparation of benzomorphans and related compounds (D. C. Palmer and M. J. Strauss, Chem. Rev., 1977, 77, 1). The following example is a recent synthesis (J. Bosch *et al.*, J. heterocyclic Chem., 1981, 18, 263).

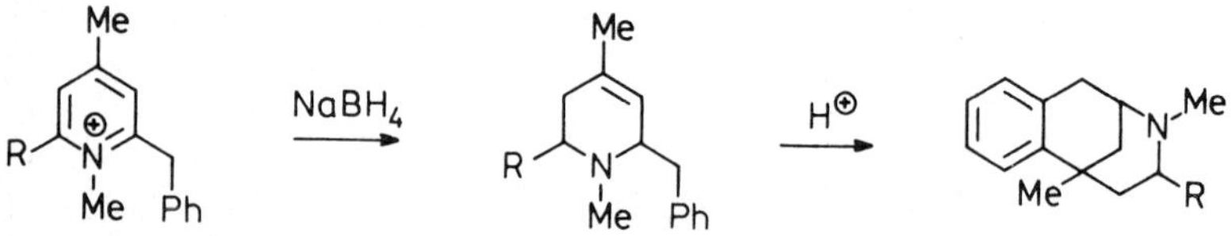

$NaBH_4$

$H^\oplus$

For other examples see F. J. Smith and G. R. Proctor, J. chem. Soc. Perkin I, 1980, 2141.

(c) Hexahydropyridines (piperidines)

(i) Preparation and reactions

Piperidines can be prepared from pyridines by reduction. However, by the use of suitable activating groups at the piperidine nitrogen it is possible to carry out nucleophilic reactions in the 2-position (D. Seebach and D. Enders, Angew. Chem., 1975, 87, 1).

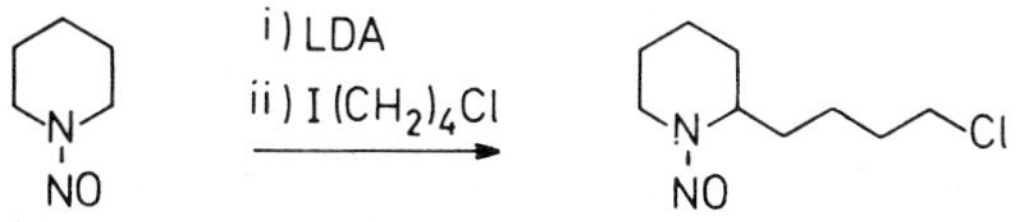

A 2-cyano group will also activate the α-hydrogen towards nucleophiles, as shown below (G. Stork, R. M. Jacobsen, and R. Levitz, Tetrahedron Letters, 1979, 771).

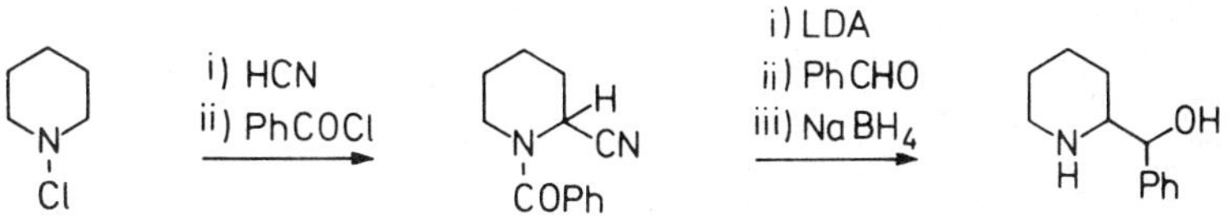

70% overall

Aliphatic aldehydes may be used instead of benzaldehyde. Another method is the use of 1-piperideine for starting material, as in the example below for the synthesis of pelletierine (J. Quick and R. Oterson, Synthesis, 1976, 745).

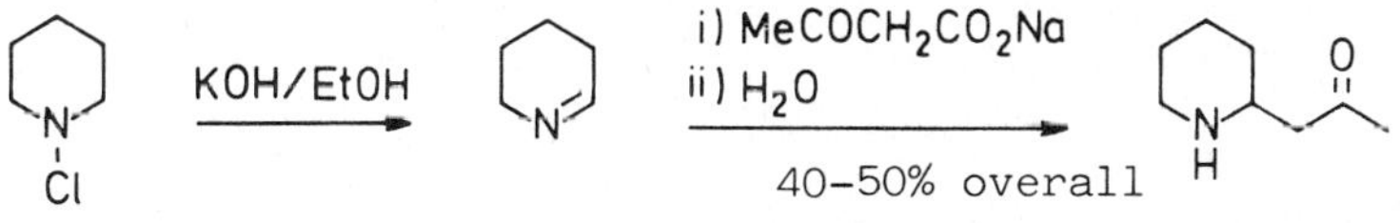

40-50% overall

A similar example is shown below for the synthesis of solenopsin starting from cyclopentanone and finally reductive alkylation of the 1-piperideine (Y. Matsumura, K. Maruoka, and H. Yamamoto, Tetrahedron Letters, 1982, 1929). Under the reaction conditions used, the stereoselectivity is > 95%.

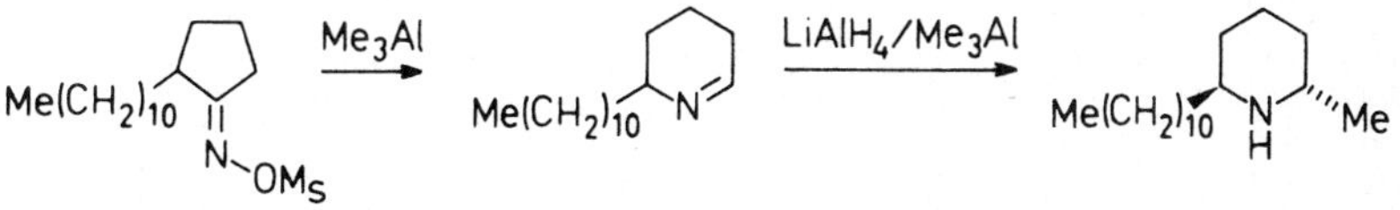

(ii) Piperidine nitroxides

An important type of piperidine derivative is the piperidine nitroxide.

For reviews see E. G. Rozantsev and V. D. Sholle, Synthesis, 1971, 190; J. F. Keana, Chem. Rev., 1978, <u>78</u>, 37. Such 2,2,6,6-tetramethylpiperidine-1-oxyls are stable radicals and therefore useful as spin-labels and spin-traps, many compounds having different R groups have been reported. Another aspect of the chemistry of these amino oxides is their redox properties. It is possible to carry out one electron oxidation of oxygen to produce the superoxide ion (T. Miyazawa, T. Endo, and M. Okawa, J. org. Chem., 1985, <u>50</u>, 5389).

$$R = OCH_2Ph$$

For other oxidations of the hydroxide ion and of alcohols see T. Endo *et al.*, J. Amer. chem. Soc., 1984, <u>106</u>, 3877; M. F. Semmelhack, C. S. Chov, and D. A. Cortes., *ibid.*, 1983, <u>105</u>, 4492. For an example of the synthesis of spin-labeled aliphatic acids see D. Gala, R. Schultz, and R. Kreilick, Canad. J. Chem., 1982, <u>60</u>, 710.

Chapter 25

BICYCLIC COMPOUNDS CONTAINING A PYRIDINE RING:
CYCLOPOLYMETHYLENEPYRIDINES, CYCLOALKENOPYRIDINES

C.D. JOHNSON

Since the publication of the 2nd Edition, not a great deal of work has been done in this area, but what there is proves of considerable interest. Comparisons and contrasts arise with alkylpyridine chemistry (Chapter 26, Section 4(b)). Thus in both types of compounds, activation of C-H bonds in conjugation with the pyridine nitrogen atom occurs.

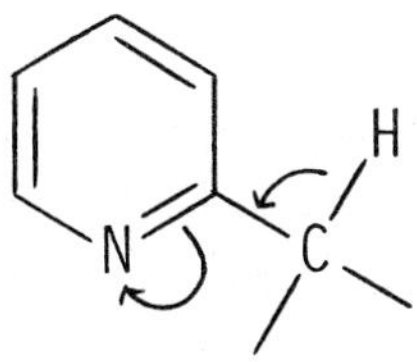

On the other hand ring strain and hybridisation changes, particularly in the smaller rings, initiate differences from alkylpyridine behaviour.

The simplest member of the series (1) does not appear to have been detected. It could potentially arise as an intermediate in the photochemical interconversion of phenylnitrene and the isomeric pyridylmethylenes, but this is not considered likely (O.L. Chapman, R.S. Sheridan and J.P. LeRoux, J. Amer. chem. Soc., 1978, 100, 6245).

1

2

3

Cyclobuta[b]pyridine (2) and cyclobuta[c]pyridine (3) have been obtained by vacuum flash photolysis of propargyl 4-pyridyl ether (J.M. Riemann and W. S. Trahanovsky, Tetrahedron Letters, 1977, 1867). The reaction is postulated to proceed *via* intermediate (4) from which the position of the nitrogen atom may be scrambled as shown.

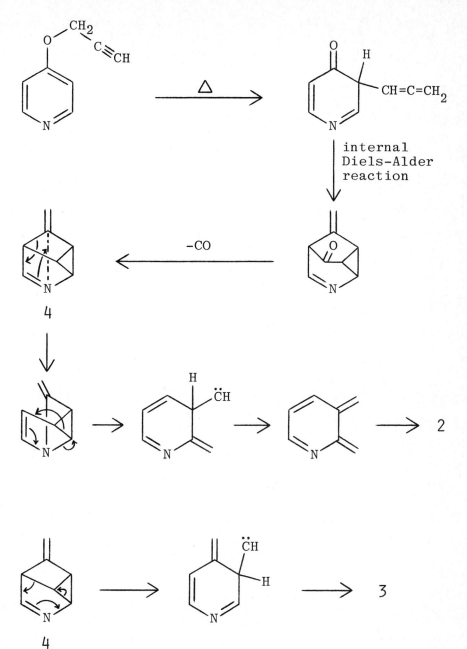

The compounds (2) and (3) can also arise from picolylcar-
bene-azatropylidene systems generated by pyrolysis of a
series of 5-(methyl-2-, -3- and -4-pyridyl)tetrazoles (W.D.
Crow, A.N. Khan and M.N. Paddon-Row, Austral. J. chem., 1975,
28, 1741).

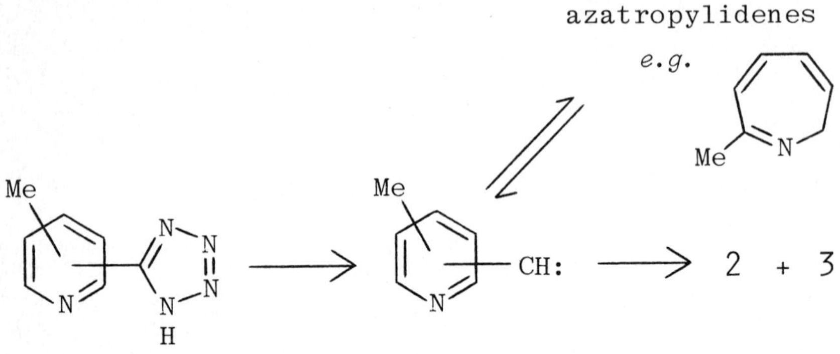

azatropylidenes

e.g.

The red pigment rubrolone (5) is a derivative of 3,4-tri-
methylenepyridine and the synthesis of such molecules has
attracted attention. A new and convenient synthesis of a
3,4-trimethylenepyridine starts from cyclopent-2-en-1-one.
The latter compound is subjected to conjugate addition by the
cuprate derived from the anion of acetonedimethylhydrazone.
The enolate produced is trapped with butyryl cyanide and the
resulting adduct then spontaneously cyclises to (6) (T.R.
Kelly and H.-t. Liu, J. Amer. chem. Soc., 1985, 107, 4998).

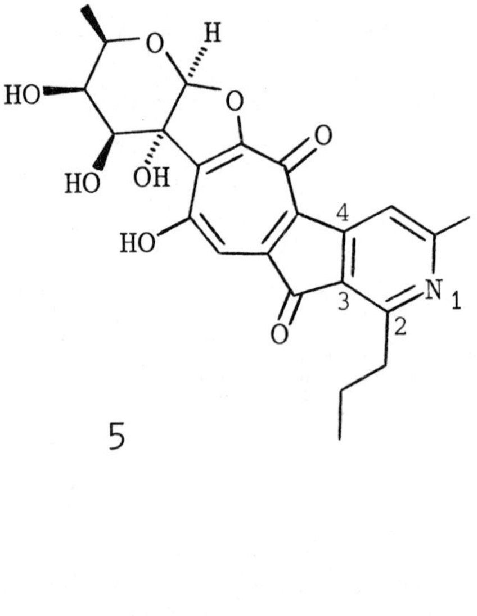

5

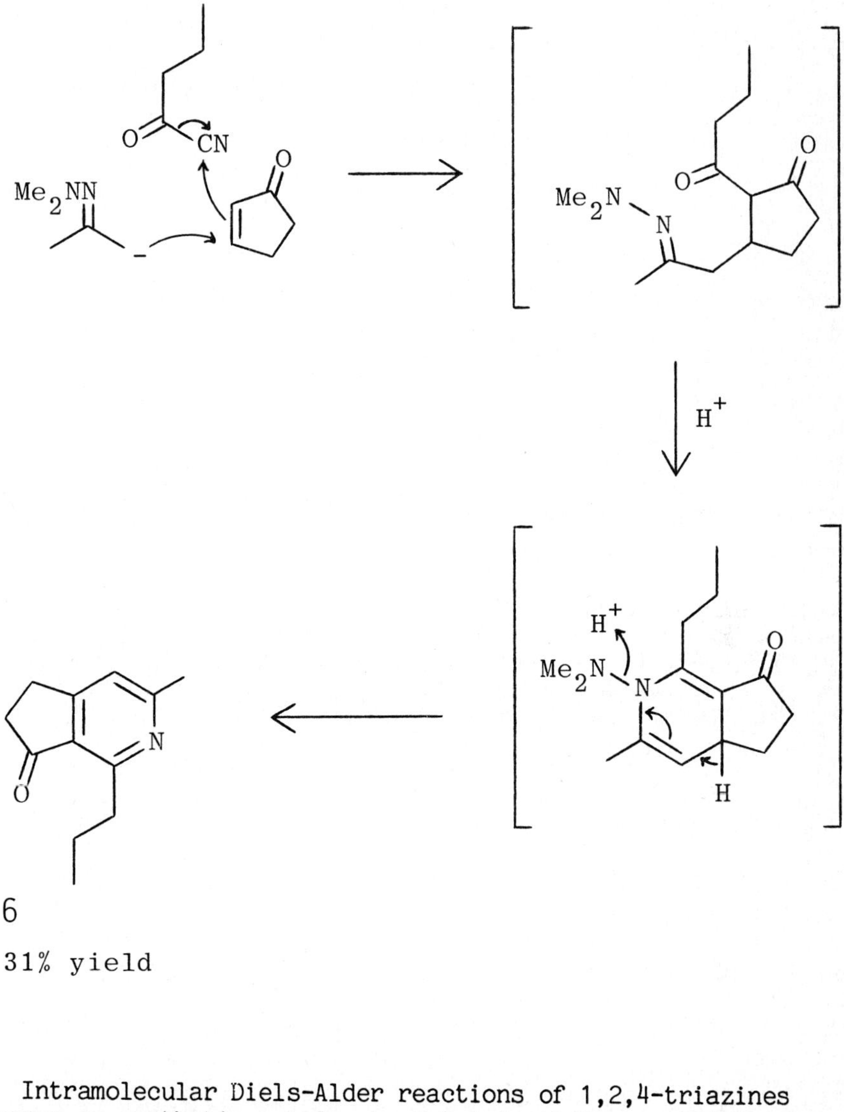

6

31% yield

Intramolecular Diels-Alder reactions of 1,2,4-triazines serve as synthetic entries to 2,3-trimethylenepyridines and 5,6,7,8-tetrahydroquinolines (E.C. Taylor and J.E. Macor, Tetrahedron Letters, 1986, 2107).

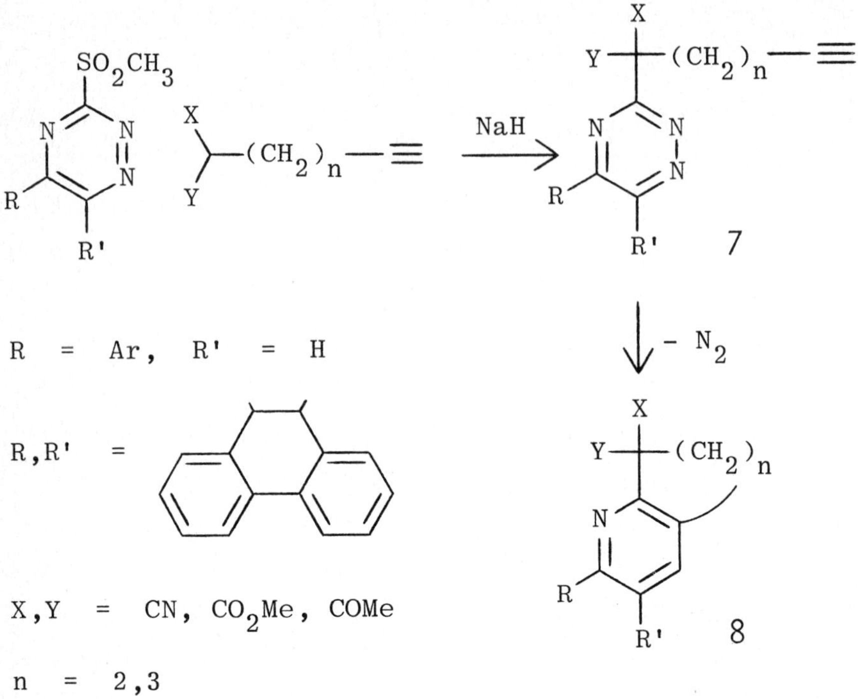
SO2CH3
X
Y
(CH2)n
NaH
7
- N2
8
R = Ar, R' = H
R,R' =
X,Y = CN, CO2Me, COMe
n = 2,3

The conversion of (7) to (8) proceeds extremely readily, even at room temperature, and yields are very high. The speed and efficiency of the reaction are ascribed to the Thorpe-Ingold (*gem*-dimethyl or scissors) effect, whereby groups X and Y stabilise the ring form (8) relative to the open chain compound (7).

Annelated pyridines are also readily prepared by condensation of cyclic ketones with the appropriate nitrogen-containing species. Friedlander condensations of β-aminoacrolein with cyclopentanone and cyclohexanone gives 2,3-trimethylene pyridine (9) and 5,6,7,8-tetrahydroquinoline (10) respectively (R.P.Thummel and D.K. Kohli, J. org. Chem., 1977, 42, 2742). These workers also prepared the bisannelated pyridines (11), (12), (13), (14), (15), (16) and (17) employing variations of the above condensation.

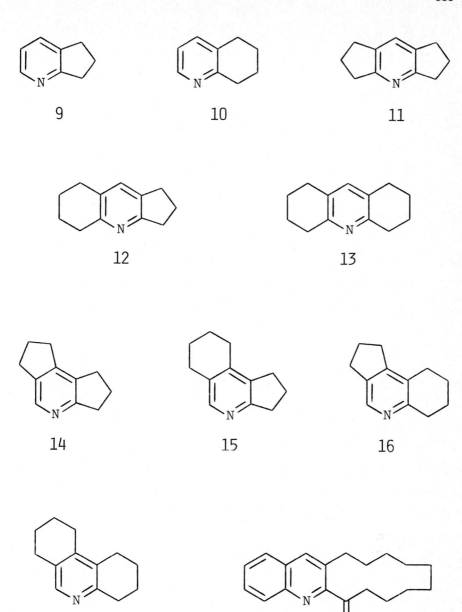

9

10

11

12

13

14

15

16

17

18

An extreme case of this type of condensation is afforded by the reaction of 2-aminobenzaldehyde with cyclododecanedione to give 8,9,10,11,12,13,14,15-octahydrocyclododeca[b]quinol-6-(7H)-one (18) (R.P. Thummel and F. Lefoulon, J. org. Chem., 1985, 50, 666).

The uv, ^{1}H- and ^{13}C-nmr spectra and pK$_a$ values of (2) and (3) and (9)-(17) have been measured (*idem. ibid.*, J. org. Chem., 1977, 42, 2742). The size and position of the annelated ring has a particularly large influence on the compounds' pK$_a$ value judged relative to the corresponding dimethylpyridine. For example, cyclobuta[b]pyridine (2) has a pK$_a$ value of 4.85 compared to 6.56 for 2,3-dimethylpyridine, while the values for cyclobuta[c]pyridine (3) and 3,4-dimethylpyridine are much closer, 6.75 and 6.61 respectively. Phenomena of this kind are interpreted in terms of rehybridisation of the atomic orbitals of the bridgehead carbon atoms used to construct the strained ring so that they have increased p-character: thus the remaining orbital has more s-character. This, in turn, means that the carbon and nitrogen atoms are bound to an orbital of higher electronegativity, which lowers the lone pair availability at N. It further results in a linear correlation between pK$_a$ and J$_{C4-H4}$ (Hz) for (11), (12) and (13) and 2,3,5.6-tetramethylpyridine.

2,3-Trimethylenepyridine (9) is also known as 6,7-dihydro-5H-1-pyrindine, expressing its correspondence to 1-pyrindine which exists as a tautomeric mixture of (19), (20) and a small amount of (21), although the latter is responsible for imparting an orange-red colour to the mixture (F. Freeman, Adv. heterocyclic Chem., 1973, 15, 187).

19   20   21

Methods for the synthesis of 5,6,7,8-tetrahydroquinoline and higher homologues have attracted much attention. Selective reduction of the benzene ring in quinoline (and isoquinoline) can be carried out in good yield by hydrogenation in concentrated hydrochloric acid over platinum oxide catalyst (F.W. Vierhapper and E.L. Eliel, J. Amer. chem. Soc., 1974, 96, 2256). Alternative reduction methods will however transform the nitrogen containing ring (R. Brettle and S.M. Shibib, (J. chem. Soc., Perkin II, 1981, 2912).

Some other convenient routes to substituted derivatives are summarised below.

(H. Mishima and H. Kurihara, Chem. Abs., 1978, 89, 163424).

(E. Ghera, Y. B. David and H. Rapoport, J. org. Chem., 1981, 46, 2059).

(K. Berg-Nielsen and L. Skattabøl, Acta. Chem. Scand., B, 1978, 553).

Chiral 5,6,7,8-tetrahydroquinolines (23) have been synthesized from (+)-(R)-pulegone (22), a popular member of the chiral pool (F. Soccolini, G. Chelucci and C. Botteghi, J. heterocyclic Chem., 1984, 21, 1001).

22

23 R = H, Me

Interest generated by the use of 5,6,7,8-tetrahydroquino-line-8-thiocarboxamides as anti-secretory agents has led to the investigation of factors governing carbophilic *vs* silicophilic attack by 8-lithio-3-methyl-5,6,7,8-tetrahydro-quinoline on silylisothiocyanates (R. Crossley and R.G. Shepherd, J. chem. Soc., Perkin II, 1985, 1917).

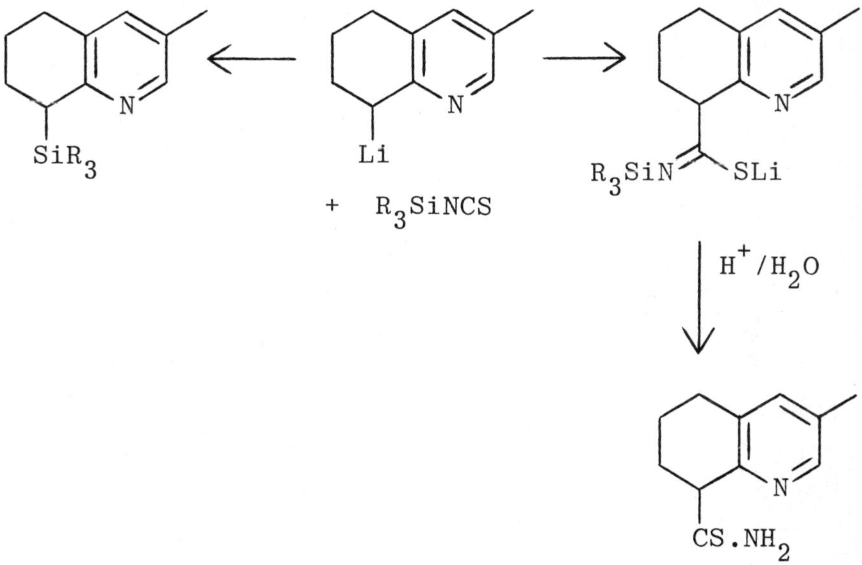

Carbophilic attack is favoured by solvents of low polarity and by bulky and/or electron donating substituents R on silicon. Treatment of 5,6,7,8-tetrahydro-2,4-diphenyl chromenylium trifluoromethane sulphonate (24) with p-toluidine gives the corresponding 1-p-tolylquinolinium salt (25) (A.R. Katritzky, L. Ürögdi and R.C. Patel, J. chem. Soc., Perkin I, 1982, 1349; A.R. Katritzky *et al.*, *ibid.*, 1985, 2159). On treatment with base, (25) yields the anhydrobase (26), which reacts readily with C-, N-, and S- electrophiles *e.g.* TsCl, PhCOCl, BrCH$_2$COOEt, PhN$_2$ *etc.*

Further examples of reacting in the annelated carbocyclic ring at the "ortho" position, activated by the adjacent sp^2-hybridised nitrogen atom, are provided by the formation of the compounds (27) (W. Dammertz and E. Rieman, Arch. Pharm., 1980, 313, 826; 1981, 314, 302).

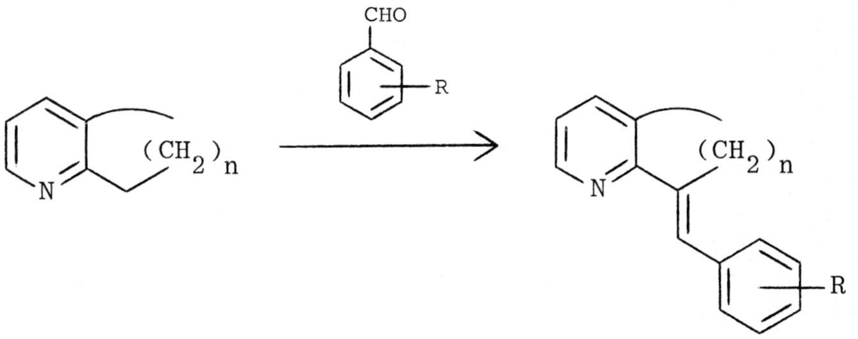

27   n = 2,3,4

R = H, *m-*,*p*-OMe

Compounds (27) are converted into benzyl compounds by catalytic hydrogenation of the exocyclic double bond. Subsequent quarternisation with methyl iodide followed by sodium borohydride reduction yields compounds (28).

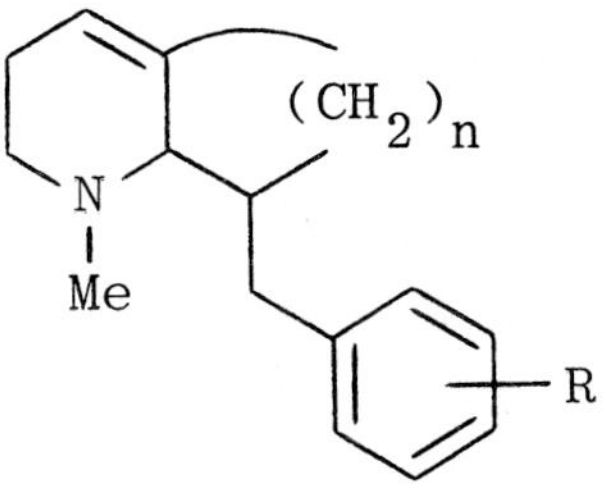

28

Chapter 26

BICYCLIC COMPOUNDS CONTAINING A PYRIDINE RING; QUINOLINE AND ITS DERIVATIVES

C.D. JOHNSON

1. Introduction

A notable contribution to accounts of quinoline chemistry is in "Comprehensive Heterocyclic Chemistry", Vol.2, ed. A.J. Boulton and A. McKillop, Pergamon, Oxford, 1984. Other important contributions are contained in "Chemistry of Heterocyclic Compounds", ed. A. Weissberger and E.C. Taylor, Wiley, New York; "Quinolines", ed. G. Jones, Vol.32, Part 1, 1977; Part 2, 1982, and in Specialist Periodical Reports of the Royal Society of Chemistry on Heterocyclic Chemistry. Advances in Heterocyclic Chemistry, ed. A.R. Katritzky, Academic Press, remains an important series frequently containing references to quinoline chemistry. Since the publication of the second edition the main thrust of research in this area appears to have been directed at synthesis, probably due to the importance of the quinoline structure in natural products, biologically active compounds and pharmaceutical products (see for example M.F. Grundon, "The Alkaloids", ed. R.H.F. Manske and R.G. A. Rodrigo, Academic Press, New York, Vol.17, 1979, p.105; H.C. Richards, Chem in Britain 1985, 21, 1001; R.V. Stevens, Acc. chem. Res., 1984, 17, 289).

2. Synthesis

The final, key, step for synthesis of the quinoline skeleton may involve formation of bond a, b, c, d or e, or process f, the annelation of the carbocyclic ring on to the appropriate pyridine.

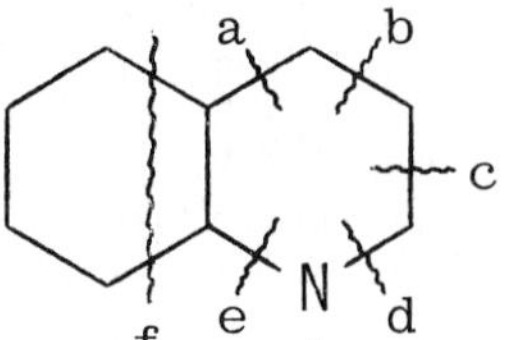

Developments in synthetic methods are considered under such headings.

(a) Formation of bond a

This includes the Skraup method and its Döbner-von Miller, von Miller-Kinkelin and Beyer modifications (G.Jones, "Heterocyclic Compounds", 1977, $\underline{32}$, 93) which proceed from aniline via intermediate (1) and subsequent acid catalysed ring closure and oxidation to yield quinolines. Conditions and procedures for these reactions have been conveniently summarised (A.G. Osborne, Tetrahedron, 1983, $\underline{39}$, 2831).

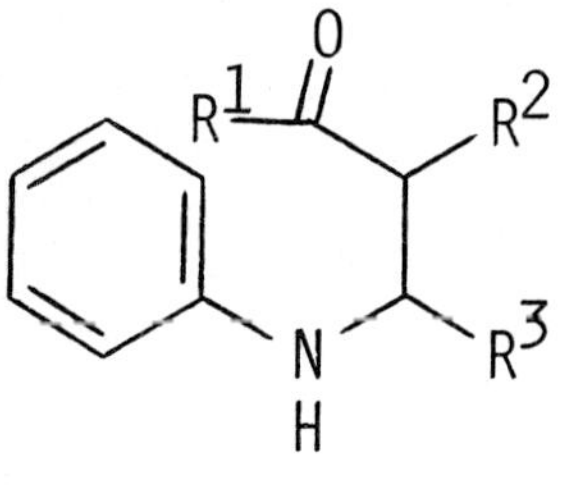

Skraup (glycerol): $R_1=R_2=R_3=H$

Skraup (methyl vinyl ketone):

$R_1=Me$, $R_2=R_2=H$

Döbner-von Miller (acetaldehyde):

$R_1=R_2=H$, $R_3=Me$

von Miller-Kinkelin (propion-aldehyde, paraformaldehyde):

$R_1=R_3=H$, $R_2=Me$

Beyer (butanone, paraformal-dehyde): $R_1=R_2=Me$, $R_3=H$; $R_1=Et$, $R_2=R_3=H$.

The Combes synthesis (G. Jones, Heterocyclic Compounds, 1977, $\underline{32}$, 119) involves initial anil formation from the reaction of aniline with pentane-2,4-dione, and does not require a final

dehydrogenation step. Related procedures leading to quinolines are the Knorr and Conrad-Limpach syntheses (G. Jones,
Heterocyclic Compounds, 1977, 32, 137) which proceeded *via* an
intermediate of β-ketoanilide or arylaminoacrylate form
respectively, from which quinolines may be obtained by
reduction, usually with phosphorus oxychloride. Steric
hindrance dictates that when *m*-toluidine and 3,4-dimethyl-
aniline are utilised in the above preparations the 7-Me
product predominates over the 5-Me product and the 6,7-Me
product over the 5,6-di-Me product, respectively (A.G.
Osborne, *loc. cit.*). This explanation is supported by the
fact that in the presence of a 4-Me group in the final
product the amount of 5- or 5,6-isomer is negligible.
Condensation of dimethyl penta-2,3-dienedioate with aniline
derivatives is also a useful route to quinolines (N.S. Nixon,
F. Scheinmann and J.L. Suschitzky, Tetrahedron Letters.,
1983, 597; J. chem. Research S, 1980, 380).

Similarly, acid-catalysed condensation of β-ketoesters with anilines affords the corresponding 2-alkyl-4-quinolones in high yield. Treatment of the latter with methyl iodide then giving mixtures of 2-alkyl-4-methoxyquinolines and 2-alkyl-N-methyl-4-quinolones (R. Somanathan and K.M. Smith, J. heterocyclic Chem., 1981, 18, 1077).

Acetanilides may be converted into 2-chloroquinoline-3-carbaldehydes in good yields by reaction with dimethyl-formamide and phosphoryl chloride (Vilsmeier's reagent) (O. Meth-Cohn and co-workers, J. chem. Soc. Perkin Trans. I, 1981, 1520, 1531, 1537, 2509).

A modification of this method utilises the reaction of an ethyl anilinobutenoate with a Vilsmeier reagent (D.R. Adams, J.N. Domínguez and J.A. Perez, Tetrahedron Letters, 1983, 517), yielding a substituted carboethoxyquinoline. Gas phase cyclisation of 1,5-diaryl-1,5-diazapentadienes gives methyl quinolines in good yield, while the extension of the method to cinnamaldehyde phenylhydrazone derivatives gives 7-substituted quinolines of high purity (H. McNab and co-workers, J. chem. Soc. Perkin Trans. I, 1984, 377, 1565, 1569)

A novel route to 2,4-diphenylquinolines is provided by the electrocyclic bicyclisation of pyrimidin-2-ones by irradiation and subsequent heating when hydrocyanic acid is lost (T. Nishio, K. Katahira and Y. Omote, Tetrahedron Letters, 1980, 2825; T Nishio and Y. Omote, J. chem. Soc., Perkin I, 1983, 1773). This constitutes a useful one-pot synthesis.

X = H, Me, OMe

Electrocyclisations are also involved in the transformation of 4H-thiazinium hydroxides to 4-quinolones (K.T. Potts, R. Ehlinger and W.M. Nichols, J. org. Chem., 1975, 40, 2596).

The initial steps of the Gould-Jacobsen (R.G. Gould and W.A. Jacobs, J. Amer. Chem. Soc., 1939, 61, 2890) synthesis are utilized in the preparation of 4-quinolone-3-carboxylic acids carrying a pyridinyl group in the 7-position (P.M. Carabateas, *et al*, J. heterocyclic. Chem., 1984, 21, 1857). The reaction proceeds through the intermediate (2). Such compounds have high antibacterial activity.

2

(b) Formation of bond b

The Friedlander synthesis is summarised below.

This reaction has been thoroughly reviewed ("Chemistry of Heterocyclic Compounds" Vol. 32, ed. G. Jones, Wiley, New York, 1977, p.181; P. Caluwe, Tetrahedron, 1980, 36, 2359; C.C. Cheng and S.Y. Yang, Org. Reactions, 1982, 28, 37) together with with Pfitzinger and von Niementowski modifications.   An improvement on the Friedlander synthesis *via* an *N*-oxide, which is subsequently reduced by phosphorus III chloride, has been described (S.B. Kadin and C.H. Lamphere, J. org. Chem., 1984, 49, 4999).

The *o*-cyano-aniline (3) gives different products depending on cyclisation conditions (H. Schaefer, K. Sattler and K. Gewald, J. prakt. Chem., 1979, 321, 695). With base (NaOMe), compound (4a) is produced, but with acid ($AlCl_3$), the acetyl group is lost to give (4b).

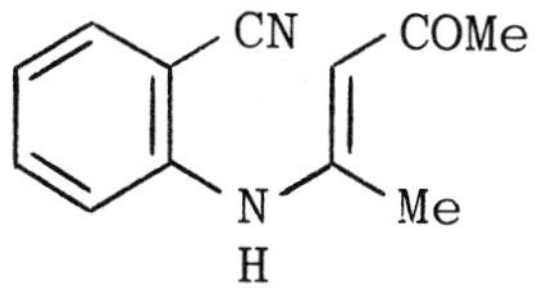

CN
COMe
N
H
Me

3

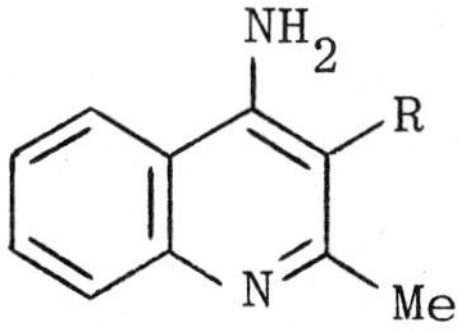

NH2
R
N
Me

4

a, R = COMe, b, R = H

A useful route to 2,4-diaminoquinolines involves base or Lewis acid induced cyclisation of *o*-amidinobenzonitriles (S.F. Campbell, J.D. Hardstone and M.J. Palmer, Tetrahedron Letters, 1984, 4813).

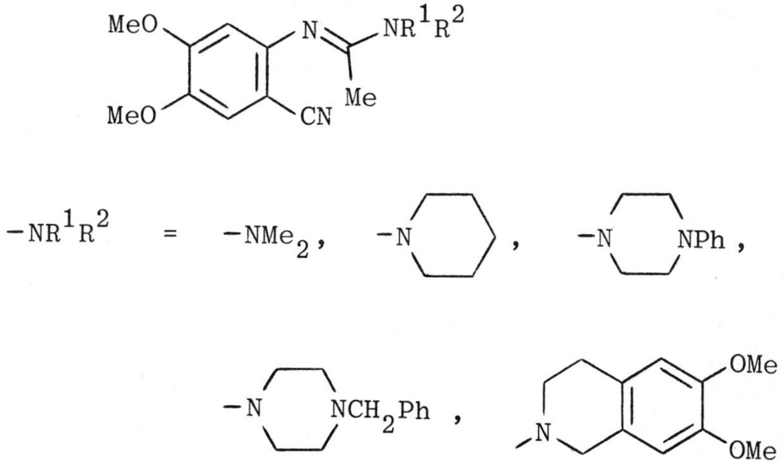

MeO
MeO
N
NR1R2
Me
CN
−NR1R2 = −NMe2,
−N
−N NPh,
−N NCH2Ph,
N
OMe
OMe

(c) Formation of bond c

This type of ring closure is comparatively rare, but cyclisation of styrenes with *o*-amino functions have been described.

$R^1 = R^2 = Me;$ $R^1 = Me, R^2 = Ph;$ $R^1 = Me, R^2 = CN;$
$R^1 = Ph, R^2 = Me$

P. de Mayo, L.K. Sydnes and G. Wenska, J. Chem. Soc., Chem. Comm., 1979, 499.

$R = CN, CHO, CO_2Me, COMe, OCOMe, Ar$

H. Horino and N. Inoue, Tetrahedron Letters, 1979, 2403.

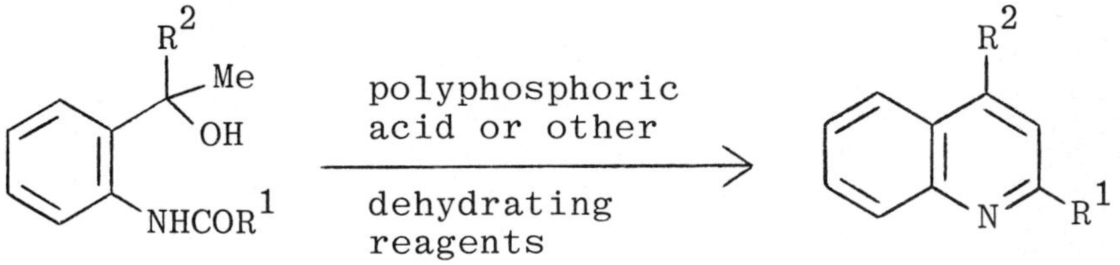

R^1, R^2 = H, Me, Ph

D.P. Curran and S.-C. Kuo, J. org. Chem., 1984, <u>49</u>, 2063.

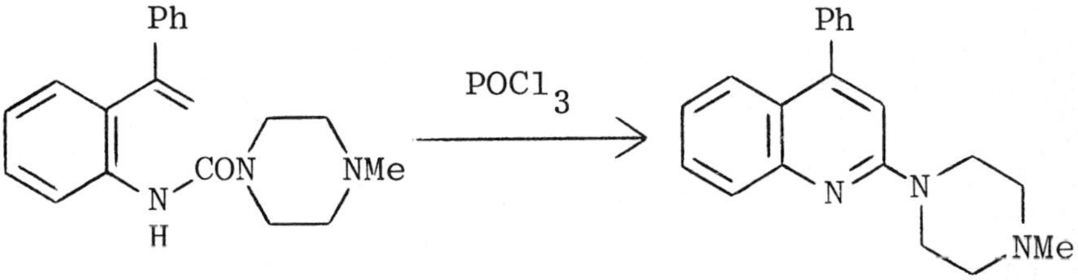

G. Gast, J. Schmutz and D. Song, Helv., 1977, <u>60</u>, 1644.

(d) Formation of bond d

o-Nitrobenzaldehyde will condense with a variety of different compounds having an active methylene group. The reaction may be seen as a modification of the Friedlander synthesis (T.L. Gilchrist, "Heterocyclic Chemistry", Pitman, London, 1985, p.272).

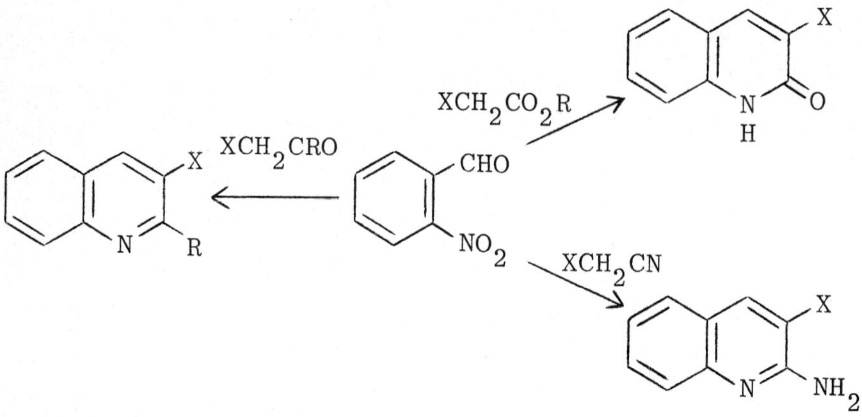

The reactions shown with appropriate reduction-cyclisation procedures have been thoroughly reviewed ("The Chemistry of Heterocyclic Compounds" Vol.32, ed. G. Jones, Wiley, New York 1977, p.207). A modification involves closure on to an epoxide (G. Jones, *op.cit*, p.220).

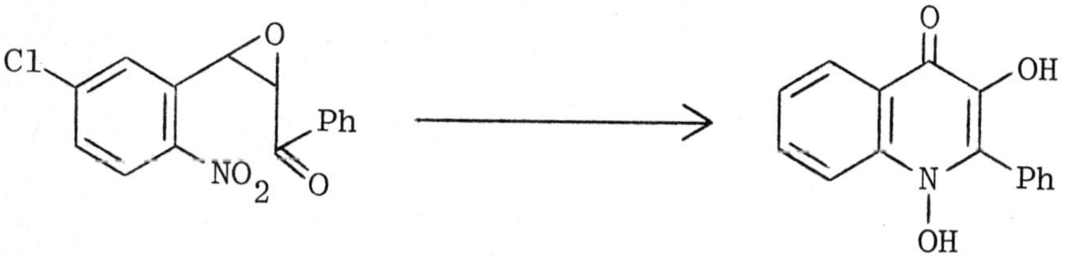

A route to quinolones involving a 2,3-sigmatropic shift (P.G. Gassman and R.L. Parton, Chem. Comm., 1977, 694) is shown below.

A Smiles rearrangement is involved in the reaction illustrated below which provides a route to the previously
inaccessible 2-acyl-3-hydroxyquinolines (D.W. Bayne, A.J.
Nichol and G. Tennant, Chem. Comm., 1975, 782).

Photoreaction of ethyl diazoacetate and diketene gives
compound (5), a valuable synthon for heterocyclic synthesis,
including, as shown, quinolones (T. Kato, N. Katagiri and R.
Sato, J. chem. Soc., Perkin I, 1979, 525).

A number of syntheses in this category from arylamine
derivatives rely on transition metal catalysis (L.S. Hegedus,
P.M. Winton and S. Varaprath, J. org. Chem., 1981, 46, 2215;
Y. Watanabe, et al., Bull. chem. Soc. Japan, 1978, 51, 3397;
N.A. Cortese, et al., J. org. Chem., 1978, 43, 2952). The
metal catalysis may occur in both making the C-C bond *ortho*
to the amino group and also in the final bond d formation
(S.E. Diamond, A. Szalkiewicz and F. Mares, J. Amer. chem.
Soc., 1979, 101, 490; L.S. Hegedus, et al., J. Amer. chem.
Soc., 1978, 100, 5800).

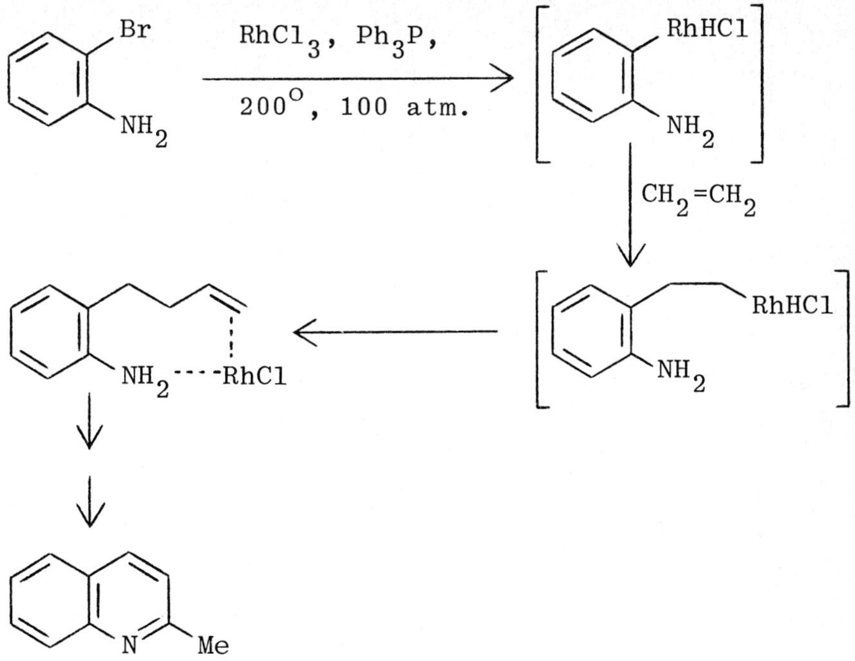
Br
NH2
RhCl3, Ph3P,
200°, 100 atm.
RhHCl
NH2
CH2=CH2
RhHCl
NH2
NH2
RhCl
N
Me

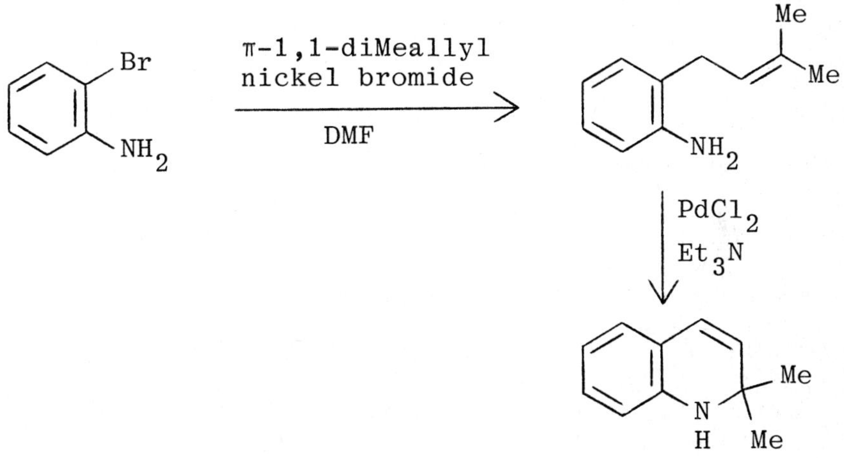
Br
NH2
π-1,1-diMeallyl
nickel bromide
DMF
Me
Me
NH2
PdCl2
Et3N
Me
N
H
Me

(e) Formation of bond e

Such syntheses, which are not normally encountered, involve intramolecular attack of an active nitrogen function, usually a radical, on an aromatic position. Examples include persulphate oxidation of (6) (A.R. Forrester, M. Gill and R.H. Thomson, Chem. Commn., 1976, 677), and photochemical irradiation of the amide (7) in the presence of iodo t-butoxide (S.A. Glover and A. Goosen, J. chem. Soc., Perkin I, 1977, 1348).

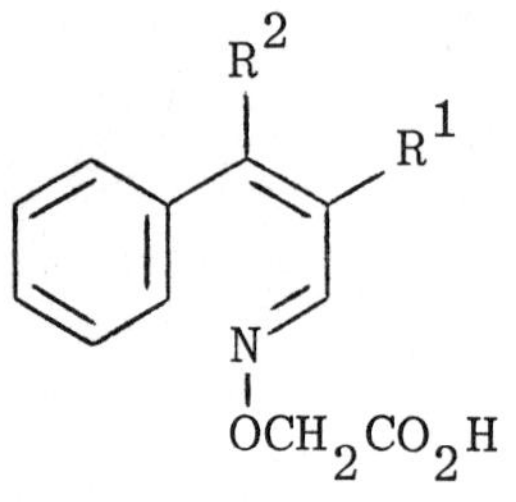

$$R^1, R^2 = H, Me$$
$$Ar = Ph, \underline{o}\text{-}MeC_6H_4$$

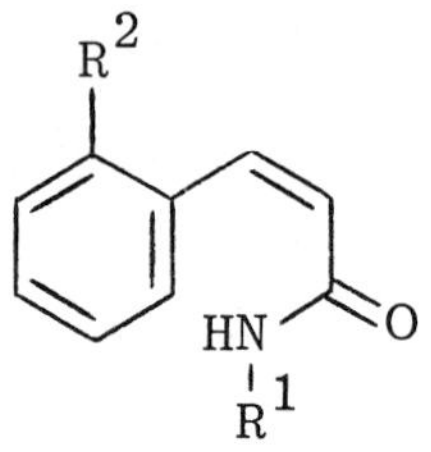

6

$$R^1 = H, Me, CH_2OBu^t$$
$$R^2 = H, Me, OMe$$

7

(f) Formation of bond f

This type of synthesis can be initiated from 2,3-disubstituted pyridines as shown for the compound (8) (A.M. van Leusen and J.W. Terpstra, Tetrahedron Letters, 1981, 22, 5097) and (9) (E.Ghera, Y. Ben-David and H. Rapoport, J. org. Chem., 1983, 48, 774).

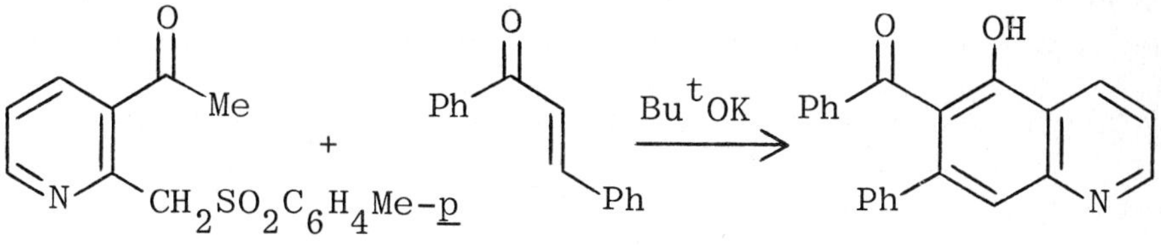

8

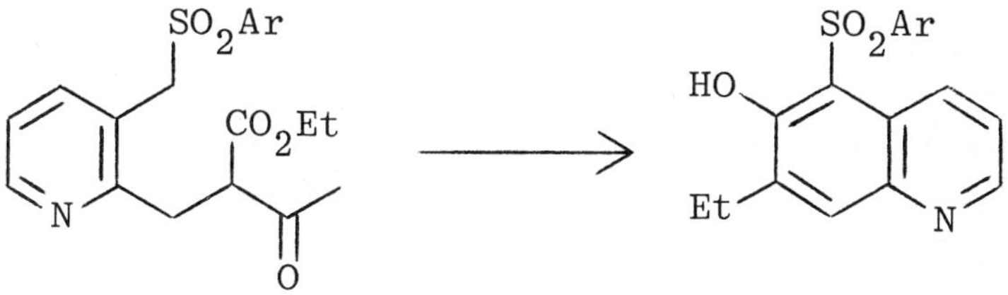

9

3. Properties of quinoline derivatives

(a) Physical and spectral properties

A good general account of tautomerism in quinolines has
been given (J. Elguero, et al., Adv. heterocyclic Chem.
Supplement 1, 1976). The ring nitrogen atom protonated form
should increase as the acidity of H in a 2- or 4-substituent
increases i.e. $OH>NH_2>CH_3$. Thus 2- and 4- quinolone exist
predominantly in the form indicated by their name, but in
certain cases (e.g. (10) in CCl_4), substituents may favour
predominance of the enol form stabilised in this case by the
through conjugation of the alkylthio-group with the ring
nitrogen atom, augmented by hydrogen bonding between the
carbonyl and hydroxyl groups.

10

Similarly, 3-acetyl-4-hydroxyquinoline exists predominantly in this form in the solid state, due to hydrogen bonding between the 4-hydroxyl and the carbonyl group of the 3-substituent: insolubility precludes study in solution (M.S. Sinsky and R.G. Bass, J. heterocyclic Chem., 1984, 21, 759).

When the potential hydroxyl group is in the benzo ring at the 5- or the 7-position, it is the hydroxy form that is favoured, and not the vinylogous amide.

The nature of the solvent will also influence the tautomeric ratio. 4-Quinolone and 3-decyl-2,8-dimethyl-4-quinolone, predominating in that form in aqueous or polar solvents, give increasing amounts of the hydroxy form as the solvent polarity, measured by Kosower's Z value, is increased (*i.e.* Z is decreased), until in cyclohexane solution, roughly equivalent proportions of both tautomers are present (J.Frank and A.R. Katritzky, J. chem. Soc., Perkin II, 1976, 1428).

The aminoquinoline tautomers are strongly favoured over the imine form; for example, log K_T = log [amine]/[imine] = 4.3 for 2-aminoquinoline. 1-Hydroxy-4-(hydroxyamino)-1,4-dihydroquinoline however is the structure preferred over 4-(hydroxyamino)quinoline-N-oxide (Y. Kawazoe, O. Ogawa and G-F. Huang, Tetrahedron, 1980, 36, 2933), considerations of which are important in the study of the carcinogenicity of 4-nitroquinoline-N-oxide (see Section 4v).

The trend is continued for the methylquinolines; for 2-methylquinoline, log K_T = log [methyl]/[methide] = 9.9 (R.A. Cox, *et al.*, Canad. J. Chem., 1976, 54, 900), estimated by a method which reveals the pK_a of the 1,2-dimethyl-quinolinium ion to be 15.23, compared with 11.68 for the 1-methyl-2-aminoquinolinium ion.

The imine-enamine tautomerism of 2- and 4-phenacylquinol-ines, together with relevant pK_a values, has been elucidated (A.R.E. Carey, *et al.*, J. chem. Soc., Perkin II, 1985, 1711).

In other basicity measurements, it is found that the nitrogen lone-pair charge of quinoline, considered as a member of a series of heteroaromatic nitrogen systems, correlates well with protonation energies derived from gas phase measurement (J. Catalán, *et al.*, J. Amer. chem. Soc., 1984, 106, 6552).

Basicities can change in the excited state of molecules; in 6- and 7-hydroxyquinoline the ring nitrogen and the hydroxyl group become more basic and more acidic respectively. In 7-hydroxyquinoline the excited state proton transfer from the neutral form to the zwitterionic tautomer has been studied by two step laser excitation fluorescence spectros-copy (M. Itoh, T. Adachi and K. Tokumura, *ibid.*, 1983, 105, 4828). Details of the fluorescence spectra of substituted

2-phenylquinoline methanols, and of their quenching by amines and sulphides (G.A. Epling and K.-Y. Lim, J. heterocyclic Chem., 1984, 21, 1205) and of quinolinium perchlorate and its quenching by electron rich olefins (U.C. Yoon, *et al.*, J. Amer. chem. Soc., 1983, 105, 1204) have been provided. On the other hand, ClO_4^- and HSO_4^-, in contrast to other anions, enhance rather than quench the fluorescence of quinolinium ions by forming excited state adducts in strong acids (S.C. Chao, J. Tretzel and F.W. Schneider, *ibid.*,1979, 101, 134).

The deuterium NMR spectrum of quinoline has been measured at natural abundance by Fourier-transform technique (H.H. Mantsch, H. Saitô and I.C.P. Smith, Prog. nucl. magn. Res. Spec., 1977, 11, 211). ^{13}C and ^1H nmr measurements of *N*-alkylmethylquinolinium salts in DMSO-d_6 reveal that the additivity parameters for the chemical shifts are quite different to those of the corresponding methyl substituted quinolines (S.R. Johns and R.I. Willing, Austral. J. Chem., 1976, 29, 1617. S.R. Johns, *et al.*, Austral. J. Chem., 1979, 32, 761) both in magnitude and sometimes in sign (J. Jaroszewska, I. Wawer and J. Oszczapowicz, Org. magn. Res., 1984, 22, 323). They are also influenced by the substituent on the nitrogen atom, and even by the nature of the anion. The influence of methyl substitution in quinoline, and also the influence of the annelated benzo-ring relative to pyridine, on the ^{15}N nmr shielding data, has been measured (L. Stefaniak, *et al.*, Org. magn. Res., 1984, 22, 201). The solvent was DMSO with 0.01 m Cr(acac)$_3$ added. Natural abundance ^{15}N-chemical shifts of *trans*-decahydroquinoline, *N*-methyl-*trans*-decahydroquinoline, and 35 alkyl-substituted NH- and NCH$_3$-*trans*-decahydroquinolines have also been measured and discussed (F.W. Vierhapper, *et al.*, J. Amer. chem. Soc., 1981, 103, 5629).

Careful ^{13}C-nmr measurements at -75°C show that 8β-*t*-butyl-*cis*-decahydroquinoline exists predominantly as the conformer with *t*-Bu axial (11) (F.W. Vierhapper, Tetrahedron Letters, 1981, 22, 5161) thus avoiding destabilising interactions between *N* and *t*-Bu in (12), which amount to >22 kJ mol^{-1}. The same conformational preference is shown by the analogous 8β-methyl compounds (F.W. Vierhapper and E.L. Eliel, J. org. Chem., 1977, 42, 51). Methylation of *N* increases the amount of axial conformer in the latter case, but not in the former, possibly due to deformation of the cyclohexane ring when *t*-Bu is axial.

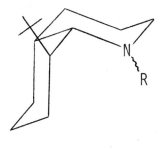
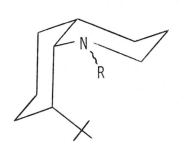

11 R = H, Me   12 R = H, Me

The [1]H- and [13]-C nmr spectra of 8-hydroxyquinoline and
5-substituted derivatives (J.Kidrić, *et al.*, Org. mag. Res.,
1981, 15, 280) and the ^{13}C-nmr spectra of 4-quinolones (A.R.
Katritzky, *et al., ibid.*,1981 16, 280) have received attention. Spin-tickling and selective population inversion
difference spectroscopy enables the sign and size of through
space [1]H-^{19}F and [13]C-^{19}F coupling constants for N-alkyl-8-
fluoroqunoini halides to be determined (M. Barfield, *et al.*,
Org. magn. Res., 1982, 20, 92). The electronic factors
influencing these couplings can be evaluated by molecular
orbital calculations of the Fermi contact term, molecular
geometrics being based on the X-ray diffraction data for
N-methyl-8-fluoroquinolinium chloride (S.R. Walter, *et al.*,
J. Cryst. spectros. Res., 1982, 12, 467). Natural abundance
^{17}O NMR spectroscopy of methyl quinoline N-oxides (D.W.
Boykin, P. Balakrishnan and A.L. Baumstark, J. Heterocycl.
Chem., 1985, 22, 1981) reveals the chemical shifts to be
particularly susceptible to electronic and steric (or
"compression") effects. the spectrum of the 8-hydroxy
compound reveals strong deshielding of [17]O-N due to hydrogen
bonding.
    By analogy with the existence of steric strain in *peri*-
substituted naphthalenes, it has been pointed out (K.
Takegoshi, *et al.*, J. org. Chem., 1985, 50, 2972) that
1,8-disubstituted quinolinium derivatives have a similar
geometry and therefore similar steric interactions exist.
Using temperature dependence measurements on the ^{1}H spin-
lattice relaxation times in the solid state of 1,8-di-

methylquinolinium iodide, the barrier to rotation of the $C_{sp}2$-CH_3 group is found to be 14 kJ mol^{-1} and that for the $N_{sp}2$-CH_3 group 10 kJ mol^{-1}.

(b) *Chemical properties: reactions*

(i) *Electrophilic substitution*

The exhaustive studies of heteroaromatic electrophilic substitution by A.R. Katritzky and his co-workers have culminated in important papers, which summarise date for nitration (J. chem. Soc., Perkin II 1975 1600) and hydrogen exchange (J. chem. Soc., Perkin II,1973, 1065; 1975, 1624; 1978, 613). The standardised results for quinoline molecules are given in Table 1; these reveal the deactivation of quinoline relative to naphthalene, and the greater intrinsic reactivity of the nitration media compared to that for hydrogen exchange. The relatively less deactivated benzofused ring is more strongly attacked in all cases, although 3-position attack may be observed, particularly with an electron donating 4-substituent.

$$E^+ \qquad E^+ \ = \ \overset{+}{N}O_2, \ D_3O^+$$

TABLE 1

Nitration and hydrogen exchange rates for quinoline and derivatives

Compound	Acid range[a]	T/°C	$\dfrac{d(\log k_{obs})}{d(-H_o)}$	Reacting Species[b]	Position (%)	$-\log k_2$[c]
Nitration						
Quinoline	77-87	25	2.19	CA	5 (50)	6.28
					8 (42)	6.36
	92-98	25	-0.60[d]	CA	5 (50)	5.86
4-quinolone	81-85	25	2.45	CA	6 (81)	4.69
					8 (19)	5.32
1-Me-4-quinolone	77-86	25	2.28	CA	3 (14)	5.67
					6 (86)	4.88
4-OMe-quinoline	81-85	25	2.50	CA	6 (72)	5.10
					8 (28)	5.51
cf						
Naphthalene	61-71	25	2.20	FB	1 (19)	-1.85
					2 (9)	-0.85
Hydrogen exchange: deuteriodeprotonation						
Quinoline	80-94	180	1.16	CA	5,6	14.96
4-OMe-quinoline	73-99	90	0.72	CA	3	9.52
6-OH-quinoline[e]	25-26	50	0.81	CA	5	4.97
Quinoline-*N*-oxide	77-86	180	0.58	CA	5,6	11.53
	64-86	180	0.56	CA	8	10.27

[a] $\%H_2SO_4$ range experimentally covered. [b] FB = free base, CA = conjugate acid. [c] Standard conditions: 25°, H_o -6.6 for nitration; 100°, pH 0 for hydrogen-exchange. [d] Profile levels out and becomes zero or slightly negative as formation of nitronium ion from nitric acid is stoichiometrically complete at high acidities. [e] For the 6- and 7- NH_2-quinolone, reaction occurs at 35° in pH region of acidity; here

$-\log k_2 = -0.2$ and -1.3 for CA and FB forms respectively of the 6-NH_2 compound reacting at the 5-position: $-\log k_2 = -0.4$ and -3.3 for CA and FB forms respectively of the 7-NH_2 compound at the 8-position.

Quinoline and derivatives duly take their appointed place in the computer-based algorithms for prediction of reactivity in electrophilic aromatic substitutions (M.G. Bures, B.L. Roos-Kozel and W.L. Jorgensen, J. org. Chem., 1985, 50, 4490). Overall electrophilic substitution may be effected by formation of quinolyl carbanions from halogenoquinolines and their subsequent reaction with electrophiles: thus a number of 8-substituted quinolines (8-substituent = CDO, CHO, $CH(OH)CH_3$, $CH_2CH=CH_2$, PPh_2, CH_3, $Sn(CH_3)_3$, CH_2CH_2OH) have been prepared from the lithio derivative obtained from 8-bromoquinoline and *sec*-butyl lithium (J.W. Suggs and G.D.N. Pearson, J. org. Chem., 1980, 45, 1514). Methods for making such lithium carbanions from fluoroquinolines have been elaborated (F. Marsais, E. Bouley and G. Queguiner, J. organometallic. Chem., 1979, 171, 273).

(ii) Nucleophilic substitution and addition
The presence of the electronegative nitrogen atom in the quinoline nucleus will clearly render the hetero-ring susceptible to nucleophilic attack. The subsequent course of the reaction - substitution or addition - will depend on the availability or otherwise, of potential anionic leaving groups.
Careful studies utilizing proton nmr-spectroscopy (J. Soltewicz, *et al.*, J. org. Chem., 1973, 38, 1947), backed by theoretical calculations (M. Hirota, *et al.*, Bull. chem. Soc. Japan, 1979, 52, 1498), on the best known of such reactions, the Chichibabin reaction, show that the 4-position is attacked as well as the 2-position to give compounds (13) and (14).

13   14

The ratio of (13) to (14) is 75:25 at -45°, but on warming (13) is converted into(14) irreversibly, so that 4-aminoquinoline becomes the major product. The formation of the 2-adduct is therefore kinetically controlled; the 4-adduct is the more thermodynamically stable one. Addition of potassium permanganate to the reaction mixture of potassium amide in liquid ammonia with quinoline has also been studied (H. Tondys, H.C. van der Plas and M. Wozniak, J. heterocyclic Chem., 1985, 22, 353): thus addition at -65° gives 2-aminoquinoline, at 15°, 4-aminoquinoline. Without the presence of potassium amide, $i.e.$ with reagent $KMnO_4$/liq NH_3, 3-nitroquinoline gives 4-amino-3-nitroquinoline at -40°, while 4-nitroquinoline gives the 4-nitro-3-amino compound. This latter reaction is the first instance of addition $meta$ to the ring nitrogen, activation being achieved by the nitro-substituent. Similarly, for the reaction of 4-chloro-3-quinoline, nitroactivation of the halogen substituent produces exclusive formation of 4-amino-3-nitroquinoline, and no oxidative amination at position 2 can be observed.

The control of amide addition by both thermodynamic and kinetic factors is reminiscent of the situation encountered in cyanide addition to N-methylquinolinium ion (A.I. Matern, E.O. Sidorov and O.N. Chupakhin, Zh. org. Khim., 1980, 16, 671). Reaction occurs at position 2 at -70° to -30°, but above 20° exclusively at position 4. Although 2- and 4-chloroquinolines and 4-chloroquinoline N-oxide give the corresponding quinolyl hydroxylamines on treatment with NH_2OH, 7-amino-5-chloro8-hydroxyquinoline gives the oxidised nitroso-compound which exists as a tautomeric mixture (J. H. Musser, Heterocycles, 1984, 22 1505).

Nucleophilic displacement of the 2-chloro-substituent in (15) can lead on to formation of quinolines with fused azacycles (R. Hayes and O. Meth-Cohn, Tetrahedron Letters., 1982, 1613).

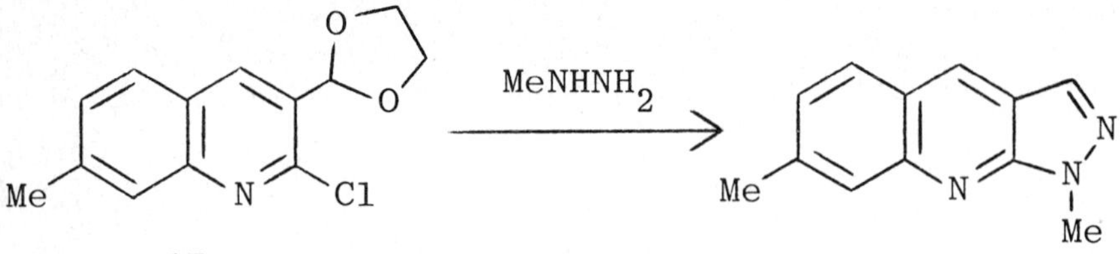

15

This type of reaction requires the aldehyde function to be converted to an acetal derivative: if however the aldehyde is transformed by Grignard reaction followed by oxidation to a ketone, the initial nucleophilic substitution is followed by facile condensation.   The chloro-substitutent in 4-chloro-1-methyl-3-carboethoxy-2-quinolone is similarly reactive (G.M. Coppola and G.E. Hardtmann, J. heterocyclic. Chem., 1981, 18, 917); for example, with methylhydrazine it gives the tricycle (16).

16

The high propensity of 8-nitroquinoline when treated with AcOH/H_2O_2 to nucleophilic addition of peracetate at the 2-position initiates a reaction sequence leading to 7-nitro-indole, among other products, the subsequent step being epoxidation of the 3,4-double bond of the 1,2-adduct (T. Kaiya and Y. Kawazoe, Tetrahedron, 1985, <u>41</u>, 511).

Nucleophilic attack of methoxide ion on the cyclic AMP phosphodiesterase inhibitor (17) yields benzyl 3-(2-methoxy-quinolyl) ketone (18) by the route shown (E.A. Harrison and K.C. Rice, J. org. Chem., 1979, <u>44</u>, 2977).

The $S_{RN}1$ mechanism for nucleophilic substitution in 2-chloro and 2-iodochloroquinoline in liquid ammonia can be electro-chemically induced (C. Amatore, *et al.*, J. Amer. chem. Soc., 1979, <u>101</u>, 6012). Cyclic voltammetry can be used to evaluate rate constants, and it is found that the quinolyl radical (19) is only attacked by soft nucleophiles *e.g.* PhS⁻ and not by harder bases such as PhO⁻.

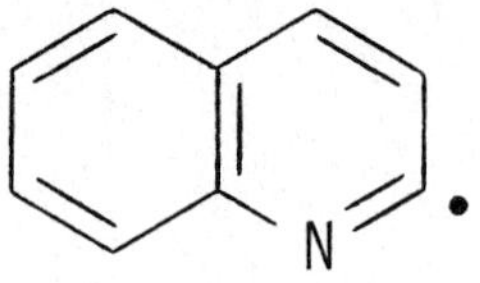

19

This quinolyl radical can also be used to abstract hydrogen atoms from alcoholates, thus yielding carbonyl compounds (C. Amatore, *et al.*, J. Amer. chem. Soc., 1982, 104, 1979).

Nucleophilic addition of hydroxyl ions to form the pseudo-bases (20) and (21) has also been extensively studied (J.W. Bunting and D.J. Norris, J. Amer. chem. Soc., 1977, 99, 1189). Compound (21) arises from the 1-methyl-5,7-dinitro-quinolinium cation here reported for the first time.

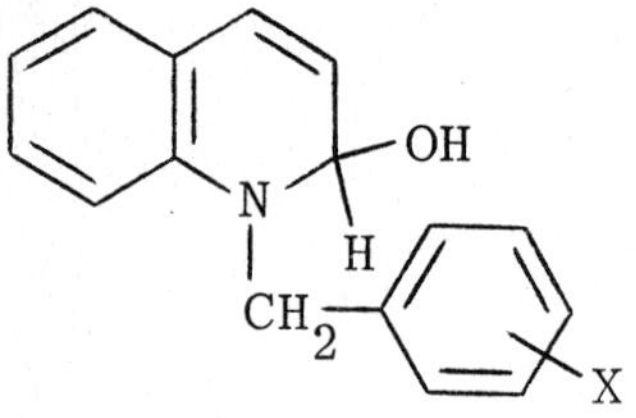

20

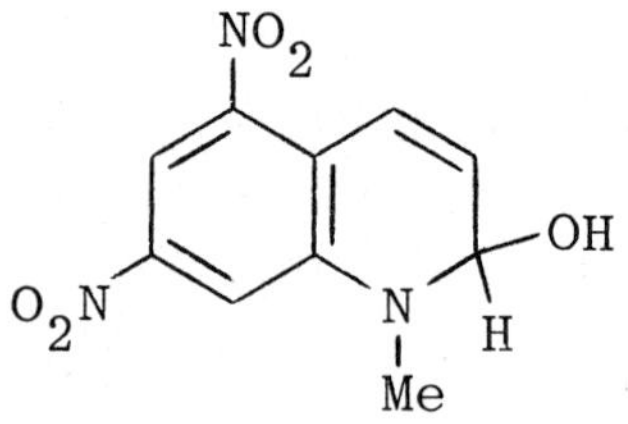

21

The pK_R^+ and k_{OH} values for formation of (20) (the latter evaluated by stopped-flow techniques) follow the equations $pK_R^+ = -1.32\sigma_X + 11.38$, and $\log k_{OH} = 0.58\sigma_X + 0.490$, revealing the ability of an electron withdrawing group X to facilitate kinetic attack adjacent to the positive nitrogen atom, and to stabilise thermodynamically the resultant product.

(iii) Free radical substitution

Italian workers justifiably contend that free radical substitution has become of major importance in heteroaromatic substitution processes (F. Minisci, Top. curr. Chem., 1976, 62, 1; E. Vismara, Chim. Ind. (Milan), 1983, 65, 34; F. Minisci, A. Citterio and C. Giordana, Acc. chem. Res., 1985, 16, 27). Thus protonated quinolines are alkylated with positional selectivity in high yields by alkyl iodides and benzoyl peroxide in acetonitrile at 80°.

$$C_9H_7{}^+NH + RI + (PhCOO)_2 \rightarrow RC_9H_6{}^+NH + PhI + PhCOOH + CO_2$$

Iron (III) salts accelerate the reaction. Some characteristic results are: 2-methylquinoline/cyclohexyl iodide, 4-substitution 92% conversion, 88% yield; 4-methylquinoline/ -butyl iodide, 2-substitution 76% conversion, 88% yield.

Similarly, the quinolinium ion with cyclohexene and silver nitratein aqueous solution in the presence of peroxydisulphate (which gives $SO_4{}^-$, a very effective electron-transfer oxidising agent) yields (22) and its 4-substituted isomer (A. Clerici, *et al.*, Tetrahedron Letters, 1978, 1149).

22

Another radical substitution process for alkylating quinolinium ions uses silver-catalysed oxidative decarboxylation of carboxylic acids (A. Citterio, V. Franchi and F. Minisci, J. org. Chem., 1980, 45 4252; F. Minisci, E. Vismara and U. Romano, Tetrahedron Letters, 1985, 4803). Peroxydisulphate, aroyl peroxides, percarbonates, or perborate may be used in this reaction, which gives high yields with good positional selectivity. The influence of solvents on free radical substitutions in 4-methyl- and 4-cyano- quinolines has also been investigated (G. Palla, Tetrahedron, 1981, <u>37</u>, 2917). The selectivity of attack by the α-amidoalkyl ·CH_2-NCH_3.CHO and the α-oxyalkyl O O radicals is particularly sensitive to solvent. In dioxane/ water both the methyl- and cyano-compounds react readily, but as formic acid is added yields decrease and the cyano-compound becomes much more reactive than the methyl-compound. It is conjectured that this is due to suppression of canonical form (23) in the resonance hybrid of the transition state.

<u>23</u>

(iv) Cycloadditions
    Reaction of *N-p*-toluene sulphonyl imino ylide derivatives of quinoline with dimethyl acetylenedicarboxylate (DMAD) yields cycloadduct (24) which eliminates *p*-toluenesulphinic acid to give pyrazolo[1,5-a]quinolines (R. Sundberg and J.E. Ellis, J. heterocyclic. Chem., 1982 <u>19</u>, 573).

2-Phenyl-, cyano-, or methyl-quinoline *N*-oxide with DMAD
yield the analogous cycloadduct (25) (Y. Ishiguro, *et al.*,
Heterocycles, 1983, 20, 1545): this yields the ring expanded
1-benzazepine by a 3,5-sigmatropic shift followed by electrocyclic ring opening. The former reaction is a vinylogous Cope
arrangement, and lack of appropriate stereochemistry in (25)
which would enable appropriate orbital overlap means that it
probably proceeds *via* a zwitterionic intermediate.

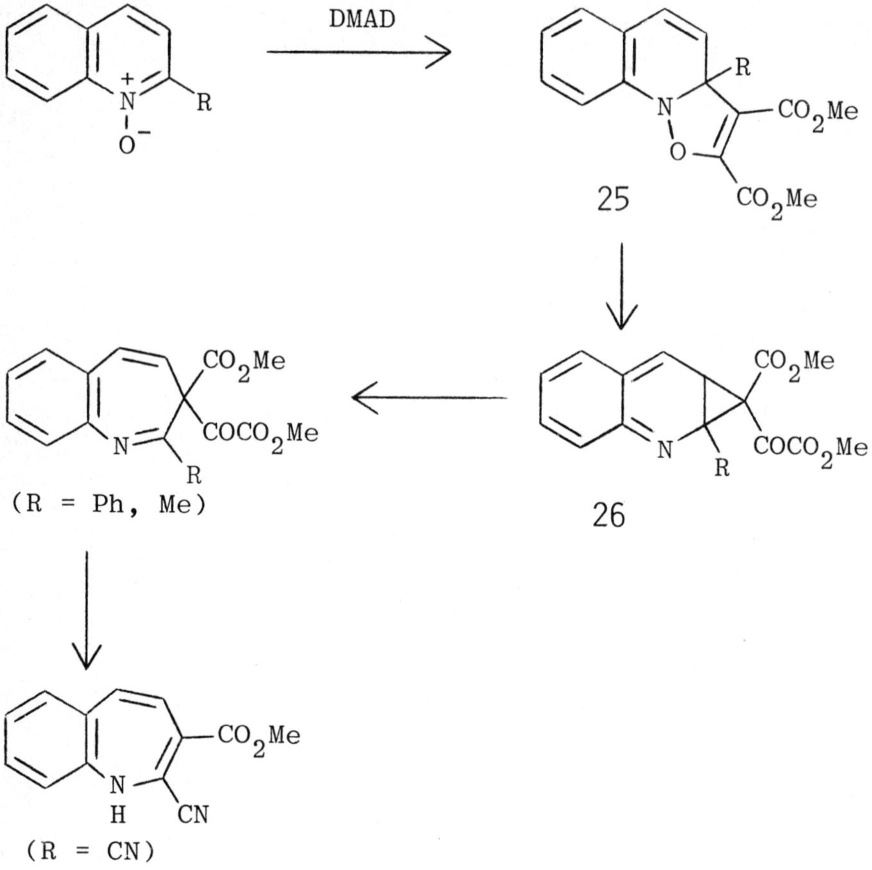

25

26

(R = Ph, Me)

(R = CN)

The 2,3-dihydroquinoline (26) is of the azanorcaradiene type, postulated as intermediates in a wide variety of 1,3-dipolar cycloadditions of quinoline *N*-oxides (R.A. Abramovitch and I. Shinkai, Acc. chem. Res., 1976, 9, 192. Photo reaction of diketene with 2-quinolone derivatives yields, *via* the spiroadduct shown (which exists as a pair of stereoisomers), the tricyclic derivative (27) (T. Chiba, M. Okada and T. Kato, J. heterocyclic. Chem., 1982, 19, 1521).

27

An important Diels-Alder adduct is formed from a quinolone methide and *N*-methylflindersine (M.F. Grundon, V.N. Ramachandran and B.M. Sloan, Tetrahedron Letters, 1981 22, 3105).

This regio- and stereospecific reaction serves to illus-
trate the potential of *N*-methylflindersine as an dienophile
in the formation of dimeric alkaloids of biological importance, for which quinolone structures function as hetero-
diene.

(v) Ring expansions
The photolysis of azidoquinolines, resulting in ring
expansion, has been extensively studied by Scriven and
Suschitzky and their co-workers. Thus methoxyazidoquinolines
yield methoxypyridoazepines (28) with primary as well as
secondary amines (Z.U. Khan, *et al.*, J. chem. Soc., Perkin I,
1983, 2495). o-Diamines (29) can also be produced. When the
added amine is cyclohexylamine (Nu = $C_6H_{11}NH-$), yields are
67% azepine, 25% o-diamine, and 4% 8-amino-6-methoxyquino-
line.

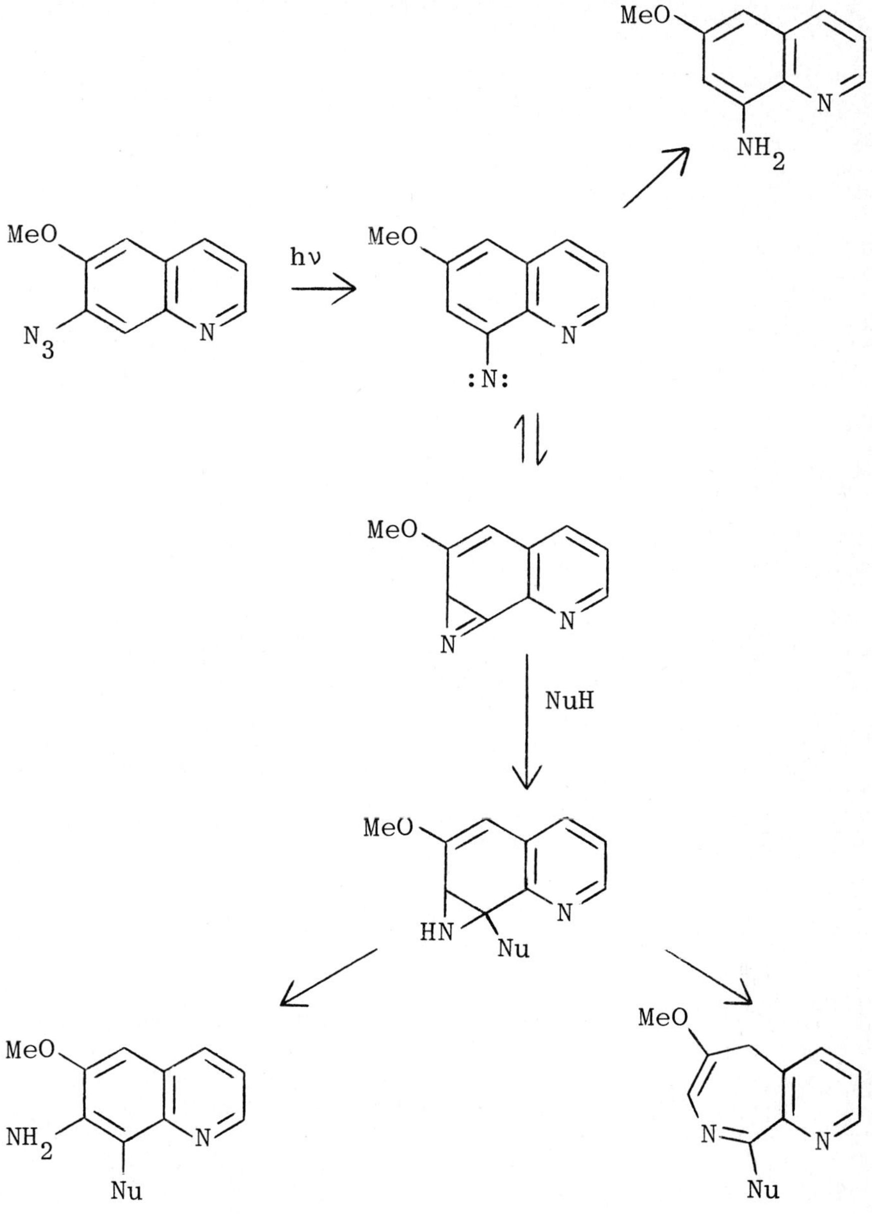
MeO
N
NH2
hν
N3
:N:
NuH
HN
Nu

Similarly, photolysis of quinolylazides in primary aliphatic amines yield bicyclic azepines, together with quinolyl--
diamines depending on conditions (F. Hollywood, *et al.*, J.
chem. Soc., Perkin I, 1982, 421).   However, secondary
aliphatic amines in this case give predominantly o-diamines.
Thus 8-quinolylazide photolysed in tetramethylethylenediamine
with a large excess of cyclohexylamine gives 57% azepine
(28), 27% o-diamine (29), and 6% 8-quinolylamine.

28   29

With diethylamine, the product is the *o*-diamine (86%
yield), and no ring expansion occurs.
Photolysis of quinolylazide in the presence of methoxide
ions affords a useful preparation of the corresponding
methoxy-substituted bicyclic azepines and benzodiazepines.
(F. Hollywood, *et al.*, J. chem. Soc., Perkin II, 1982, 431.)
The methoxy group can be replaced by amines.

However, o-alkylthioquinolylamines rather than ring expanded products result when 5- or 8-quinolylazides are photolysed in thiols (Z.U. Khan, et al., J. chem. Soc, Perkin I, 1982, 671). This work of the "Salford Syndicate" has been well summarised (P.A.S. Smith, "Azides and Nitrenes", ed. E.F.V. Scriven, Academic Press, Orlando, 1984, p.137).

In like manner, quinoline N-imides substituted in the 6- or the 8-position with an electron donating substituent yield 3H-1,3-benzodiazepines (T. Tsucheya, et al., Chem. Comm., 1981, 211).

The influence of solvent and substituents on the ring expansion of quinoline N-oxides on photolysis giving compound of type (30) has been thoroughly reviewed (F. Bellamy and J. Streith, Heterocycles, 1976, 4, 1391).

30

4. Substituted quinolines

(a) Quantitative measurements of side chain reactivity
In an extension of the accepted Hammett methodology (C.D. Johnson, "The Hammett Equation", Cambridge University Press, 1973), the nitrogen atom in quinoline may be considered as an aza "substituent" in the naphthyl system, and its electronic influence on the various ring positions measured by side

chain reactivity. Two such series have been studied (M. Sawada, *et al.*, Tetrahedron Letters., 1980, 4917; 1981, 4733).

Reaction (1)

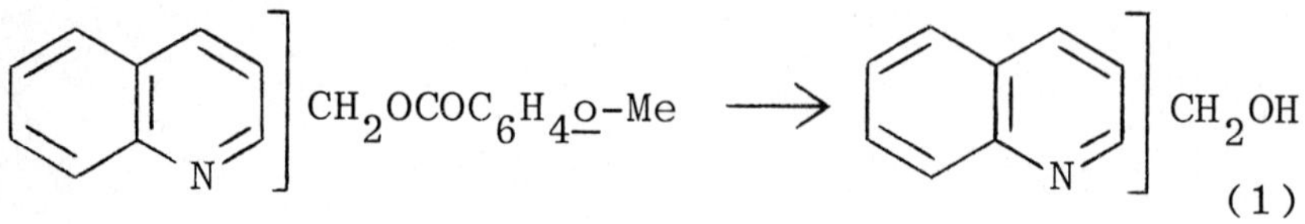

70% aqueous acetone, 25°

Reaction (2)

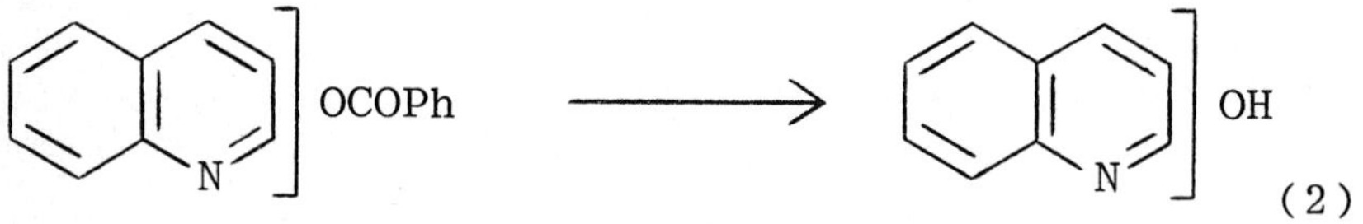

56% aqueous acetone, 5°
33% aqueous MeCN, 25°

For reaction (1), ρ is 0.989; for reaction (2), the reaction site is not "insulated" from the ring system and the relevant equations become

$$\log k/k_o = 1.93 \, (\sigma^o + 0.25\Delta\sigma_R) - 0.04$$

for the water-acetone medium, and

$$\log k/k_o = 1.35 \, (\sigma^o + 0.18\Delta\sigma_R) - 0.03$$

for the water-acetonitrile system.
The derived σ values are given below
Reaction (1), σ^o values

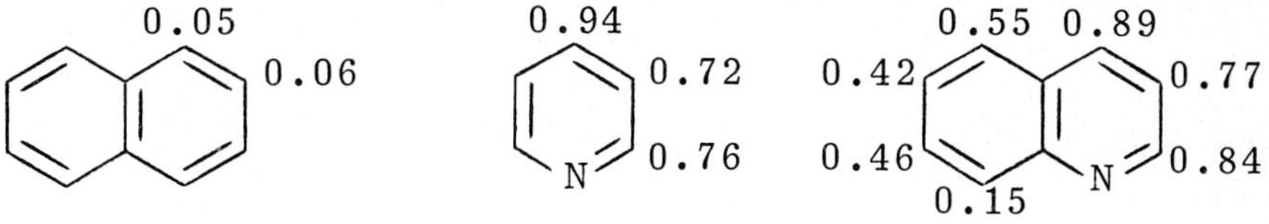

0.05
0.06

0.94
0.72
0.76

P

0.55 0.89
0.42
0.77
0.46
0.84
0.15

Q

Reaction (2), σ^- values:

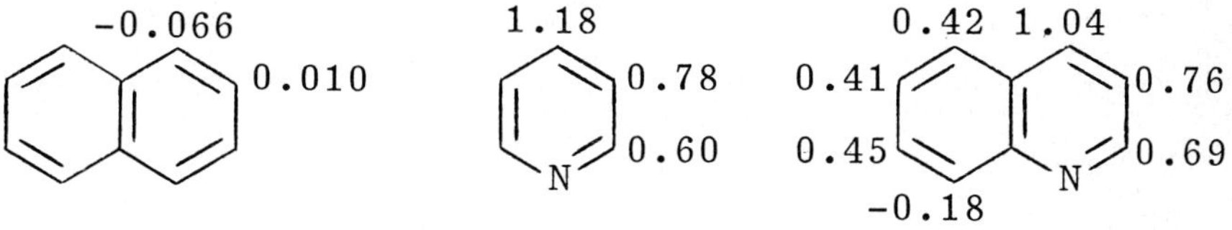

-0.066
0.010

1.18
0.78
0.60

P

0.42 1.04
0.41
0.76
0.45
0.69
-0.18

Q

In consequence, the $\bar{\sigma} - \sigma^o$ values become

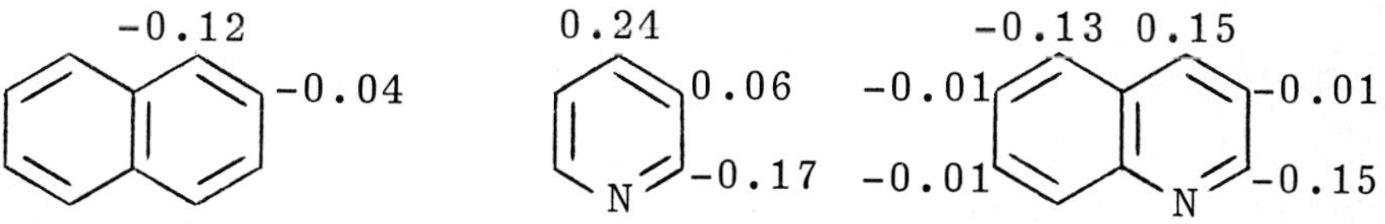

-0.12
-0.04

0.24
0.06
-0.17

P

-0.13 0.15
-0.01
-0.01
-0.01
-0.15

Q

The order of σ^o values is given below
    4Q> 2Q> 3Q> 5Q> 7Q ≅ 6Q> 8Q> Ph
    4P> 2P> 3P>   > Ph

The electron withdrawing capacity of the aza substituent is evident at all positions. Also apparent is the increased electron withdrawal from the positions (2,4,5,7) conjugated with the aza substituent and the moderated effect in the benzo-ring. The same effects are observed for reaction (2) with the main difference that the absence of the insulating methylene group in the side chain produces firstly a rate retardation in positions adjacent to N and to *peri*-hydrogen atoms, and secondly a large rate acceleration due to conjugation with 4P and 4Q positions, particularly obvious when the decelerative steric effects of 4Q are allowed for. This must be due to the importance of canonical forms reflecting the stability of 4-pyridone and 4-quinoline structures.

In another study (M. Sawada, et al., Bull. chem. Soc. Japan, 1980, $\underline{53}$, 2055; Tetrahedron Letters, 1980, 4913), the nitrogen atom is considered as a potential reaction site and substituents varied on positions 3,4,5,6 and 7. Application of the equation:

$$\log k/k_O = \rho_I \sigma_I + \rho_\pi^+ \sigma_\pi^+ + \rho_\pi^- \sigma_\pi^-$$

gives ρ values as follows

α	ρ_I	ρ_π^+	ρ_π^-
3α	6.28	1.29	–
4α	5.11	10.01	–
5α	3.89	3.32	2.1
6α	2.96	2.08	2.45
7α	3.56	4.15	2.66

The ρ_I values indicate the decreasing influence of the inductive effect as distance of the reaction site from the position of substitution is increased (considered either through space or *via* the σ-bonds), and the importance of the π donor resonance effect in the order of $4\alpha > 7\alpha > 5\alpha > 6\alpha > 3\alpha$.

(b) Substituents linked by carbon-carbon bonds
The carbon-nitrogen double bond of quinoline confers acidity on the α-protons of a 2-alkyl enhanced if the nitrogen atom is protonated. Thus the methiodide of 2-methylquinoline yields (31) with trifluoroacetic anhydride, which on warming with benzylamine gives the monotrifluoro-acetyl compound (A.S. Bailey, *et al.*, J. chem. Soc., Perkin II, 1983, 795.

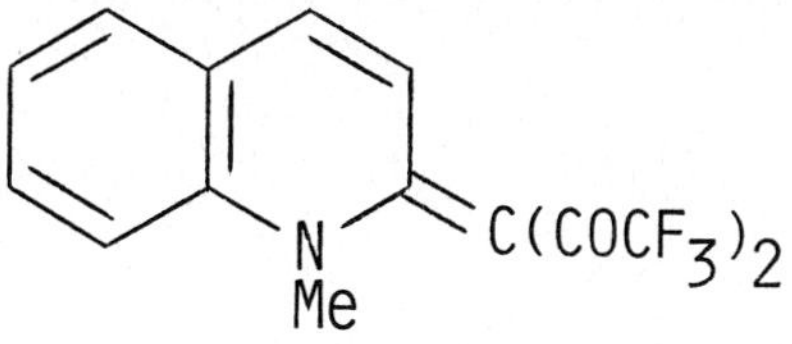

31

Likewise, 2-pyridone annelation to 2-methylquinoline-3-
carboxylic acid may be achieved by Vilsmeier reagents (O.
Meth-Cohn and H.C. Taljaara, Tetrahedron Letters, 1983,
4607), involving intermediate (32).

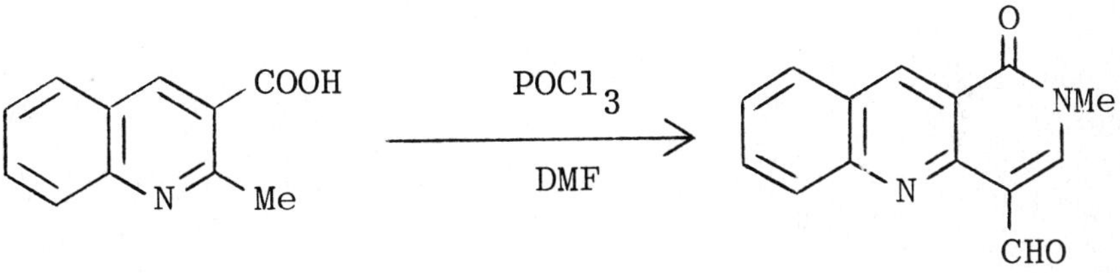

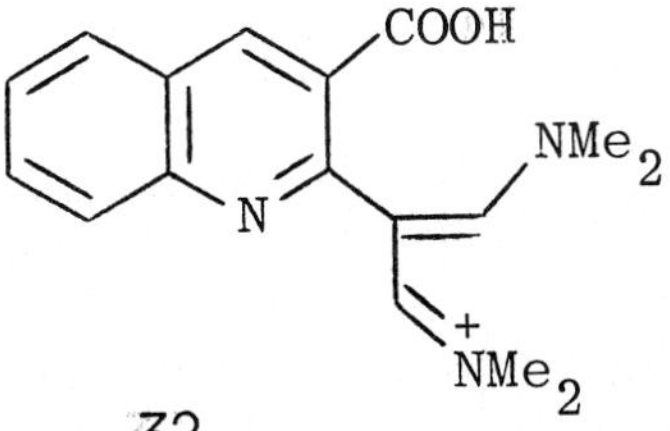

32

This acidity of the α-hydrogen atoms also play a part in the ready ene reaction of 2-methylquinoline with dimethyl acetylenedicarboxylate(DMAD), which subsequently leads on to (33) (A.M. Acheson and G . Proctor, J. chem. Soc., Perkin I, 1979, 2171).

33

2-Methylquinoline will also react with dichlorocarbene to yield spiro compound (34) which then leads on to 1,1-dialkoxy-1,2-dihydrocyclo[b]quinolines, hydrolysable in acid to 2-methylquinoline-3-carboxylic esters (Y. Hamada, M. Sugiura and H. Hirota, Tetrahedron Letters, 1981, 2893):

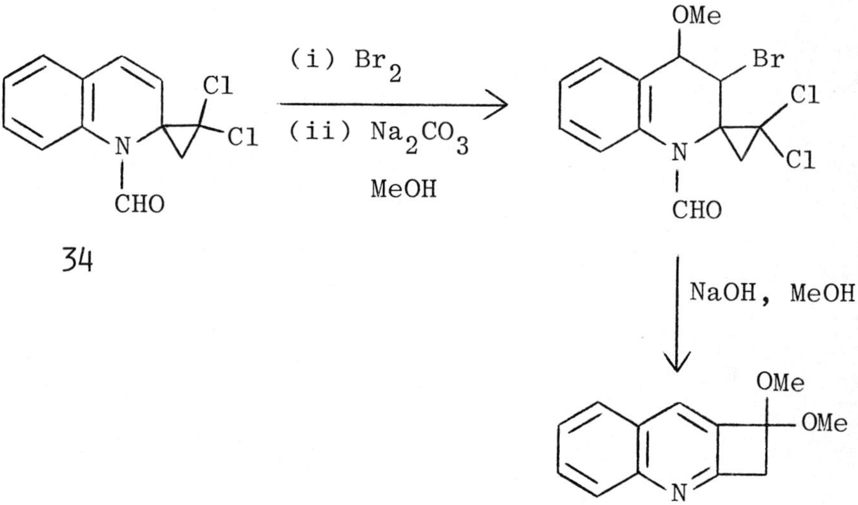

In an elegant piece of work, the introduction of a alkyl group into the 4-position of quinoline has been carried out by the reaction of quinoline with ethyl chloroformate, then trialkyl phosphite, *n*-butyllithium, and finally the appropriate alkyl halide (K.-y. Akiba, T. Kasai and M. Wada, Tetrahedron Letters, 1982, 1709). Good to excellent yields were obtained.

$$P(OR)_3$$

R = Me, Et, iso-Pr

R' = Me, Et, iso-Pr,

$CH_2=CH-CH_2-$

(i) n-BuLi

(ii) R'X

Similarly good yields of alkenylquinolines may be obtained by treatment of bromoquinolines with HC≡CR in the presence of the complex $Pd(PPh_3)_2Cl_2$, copper(II) iodide, and trimethyl-amine (H. Yamanaka, *et al.*, Chem. pharm. Bull. Japan, 1979, **27**, 270).

(c) Aminoquinolines

2- and 4-Aminoquinolines can be prepared by heating quinolones with amine hydrochloride, phosphorous pentoxide and *N*,*N*-dimethylaniline at 250° (E.B. Pedersen and D. Carlsen, Chem. Scr., 1981, **18**, 240). It is also reported that debenzylation accompanies the reaction, at 300°C, of benzyl ammonium chloride with 4-hydroxy-2-quinolones to give 4-amino-2-quinolones. (W. Stadlbauer and T. Kappe, Synthe-sis, 1981, 833). Photolysis of azidoquinolines and quinoline *N*-oxides, as well as furnishing ring expanded products (see 3(a) (v)), can also yield amino compounds *via* a radical process (H. Sawenashi, T. Harai and T. Tsuchiya, Hetero-cycles, 1984, **22**, 1501): thus 3-azidoquinoline-*N*-oxide on irradiation in acidified alcohols gives a 3-alkoxy-4-amino-quinoline-*N*-oxide. In other cases the *N*-oxide function may be lost. The preparation of quinolines substituted with diamine functionalities in the 3-, 5- or 7-position has been studied with a view to their potential as antiparasitic agents (M.S. Khan and M.P. LaMontagne, J. med. Chem., 1979, 22, 1005; A. Maskovac, *et al.*, J. heterocyclic Chem., 1982, 19, 829).

The normal site for alkylation of aminoquinolines is, as for protonation, the ring nitrogen atom. However, due to a steric effect,. 8-aminoquinoline reacts with methyl iodide at the exocyclic nitrogen (L.W. Deady and N.I. Yusoff, J. heterocyclic Chem., 1976, 13, 125). 8-Acetylaminoquinolines can be *N*-aminated and then cyclised to yield [1,2,4]-triazino[1,6,5-ji]quinolines (35) (Y. Tamura, *et al.*, Heterocycles, 1977, 61, 281). Oxidation of 8-aminoquinolines of structural form of the antimalarial drug primaquine (36) can lead to either ring or chain *N*-oxide formation (J.L. Johnson, *et al.*, J. heterocyclic Chem., 1984, 21, 1093).

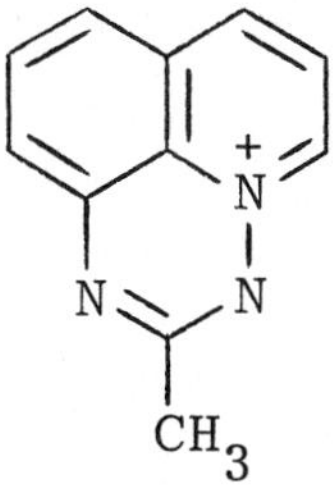

35

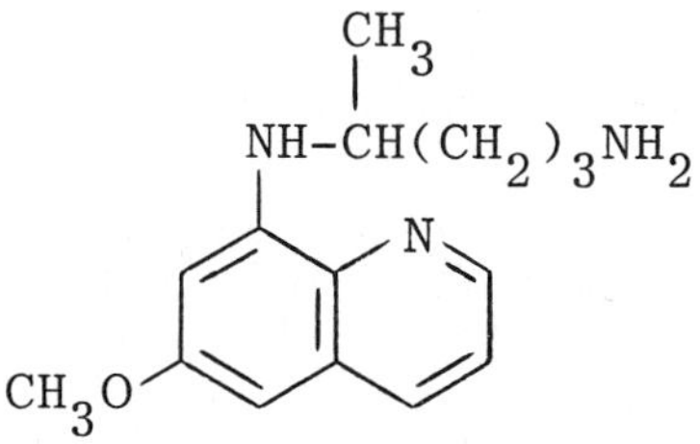

36

5-Amino-1-ethyl-6,7-methylenedioxy-4-quinolone 3-carboxylic acid, an analogue of the anti-bacterial compound, oxolinic acid, behaves anomolously on diazotisation, giving fission of the alkoxy-substituent, and produced a diazo-oxide (J. Frank, *et al.*, *ibid.*, 1981, <u>18</u>, 985).

(d) Reissert and related reactions

The Reissert reaction, comprehensively reviewed by F.D. Popp (Adv. heterocyclic Chem., 1979, <u>24</u>, 187), has been shown to be useful for the direct introduction of nitrile substituents into the quinoline nucleus (D.L. Boger, *et al.*, J. org. Chem., 1984, <u>49</u>, 4056).

A methoxy substituent in the 6-position of quinoline also sponsors the decomposition of Reissert compound (37) to 2-cyano-6-methoxyquinoline in the presence of thallium(III) nitrate and trimethyl orthoformate: if the methoxy group is in the 7-position ring contraction occurs (E.C. Taylor, I.J. Turchi and A. McKillop, Heterocycles, 1978, 11, 481).

The anion of a Reissert compound bearing a 4-position blocking group can act as a nucleophile in Michael addition to acrylonitrile to yield pyrrolo[1,2-a]quinolines (B.C. Uff, *et al.*, J. chem. Soc., Perkin I, 1977, 2018).

The Reissert-Henze reaction involves quinoline *N*-oxides, and in a variant of this, 3,4-dicyanoquinoline is produced by the reaction of 4-nitroquinoline *N*-oxide as shown below (S. Harusawa, Y. Hamada and T. Shioiri, Heterocycles, 1981, 15, 981).

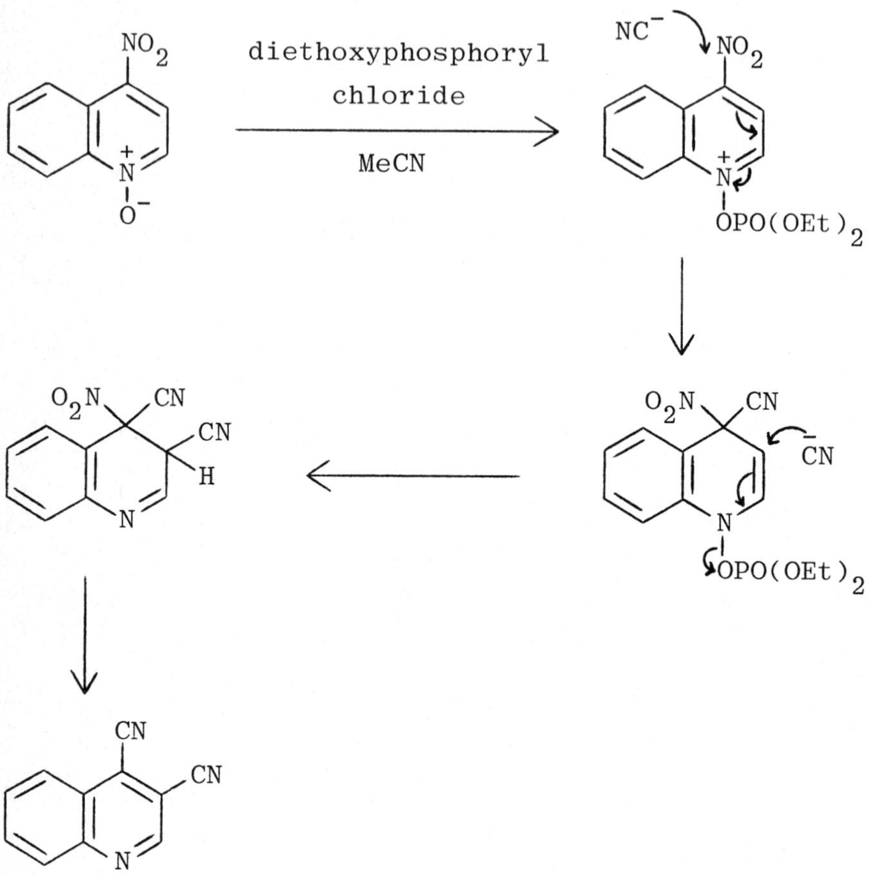

(e) Quinoline N-oxides
Quinoline *N*-oxide react readily with acetic anhydride
forming 2-quinolones, a facile method for their preparation
(S. Oae and K. Ogino, Heterocycles, 1977, 6, 583); the same
type of mechanism is involved in the conversion of quinoline
N-oxides to 2-quinolones, by treatment with *p*-toluene-
sulphonyl chlorides in aqueous base. This type of reaction of
quinoline *N*-oxides can be used to effect functionalisation of
the 2-position by treatment with enamines in the presence of
acylhalides (H. Yamanaka, H. Egawa and T. Sakamoto, Chem.
pharm. Bull. Japan, 1978, 26, 2759); thus if 5-amino-
isoxazoles are used as the enamine component, then 2-(4-
isoxazolyl)-quinolines are formed.The conversion to 2-cyano-
quinoline with diethyl phosphorocyanide (S. Harusawa, Y.
Hamada and T. Shioira, Heterocycles, 1981, 15, 981), and the
reaction with compounds bearing acidic hydrogen atoms in the
presence of acetic anhydride (M. Iwao and T. Kuraishi, J.
heterocyclic Chem., 1978, 15, 1425) also constitute examples
of this mechanistic process.

These latter products are considered to exist in the above form, presumably because the extended conjugation thus realised makes up for the partial loss of aromaticity. However, with ethyl cyanoacetate and acetic anhydride in dipolar aprotic solvents (DMF or DMSO) the initial C-adduct gives mainly the N-substituted derivative (K. Funakoshi, *et al.*, Chem. pharm. Bull. Japan, 1978, 26, 3504).

Substituents in the benzo-portion of the quinoline *N*-oxide may considerably influence such reactivity. For example, 8-methoxyquinoline *N*-oxide with acetic anhydride and methanol gives 2,8-dimethoxyquinoline, while the 6-methoxy compound with acetic anhydride and ethanol yields 2-ethoxy-6-methoxy-quinoline, but 7-methoxyquinoline *N*-oxide does not react under these conditions (M.J. Dimsdale, J. heterocyclic Chem., 1979, 16, 1209). 2-Azido-4-methylquinoline *N*-oxide undergoes ring contraction when heated (R.A. Abramovitch and B.W. Cue J. org. Chem., 1980, 45, 5316).

Interest continues on the carcinogenicity of 4-nitroquinoline
N-oxide (T. Sigimura, "Carcinogenesis." Vol.6. "The Nitro-
quinolines", Raven Press, New York, 1981). Its metabolic
transformation involves conversion into 4-hydroxyaminoquino-
line *N*-oxide (Y. Kawazoe, O. Ogawa and G.F. Huang, Tetrahedron, 1980, 36, 2933) and thence to the possible ultimate
carcinogens (38a, b and c) (M. Demeunynck, M.F. Lhomme and J.
Lhomme, Tetrahedron Letters, 1981, 3189; M. Demeunynck, *et
al.*, J. heterocyclic Chem., 1984, 21, 501.

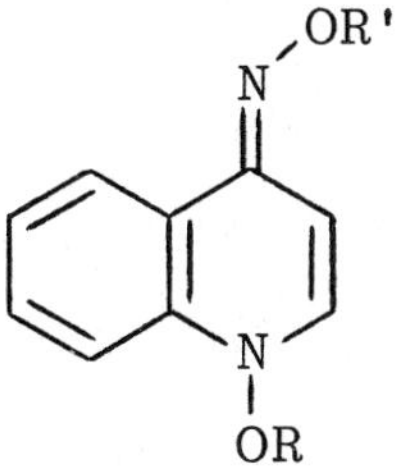

38

a;   R = R' = Ac

b;   R = Ac,   R' = H

c;   R = H,   R' = Ac

The solvolytic behaviour of (38a) has been studied (M. Demeunynck, M.F. Lhomme and J. Lhomme, J. org. Chem., 1983, 48, 1171), and also its binding to the nuclei acid base adenosine (N. Thome, *et al.*, Tetrahedron Letters, 1985, 26, 3799).

(f) Quinolones
   Facinating reactions occur when the oxindole-acrylate (39) is treated with diazomethane (G.B. Bennett, R.B. Mason and M.J. Shapiro, J. org. Chem., 1978, 43, 4383). Ring expansion results to give a 2-quinolone, which then slowly equilibrates in DMSO with the tricyclic furano-compound (40).

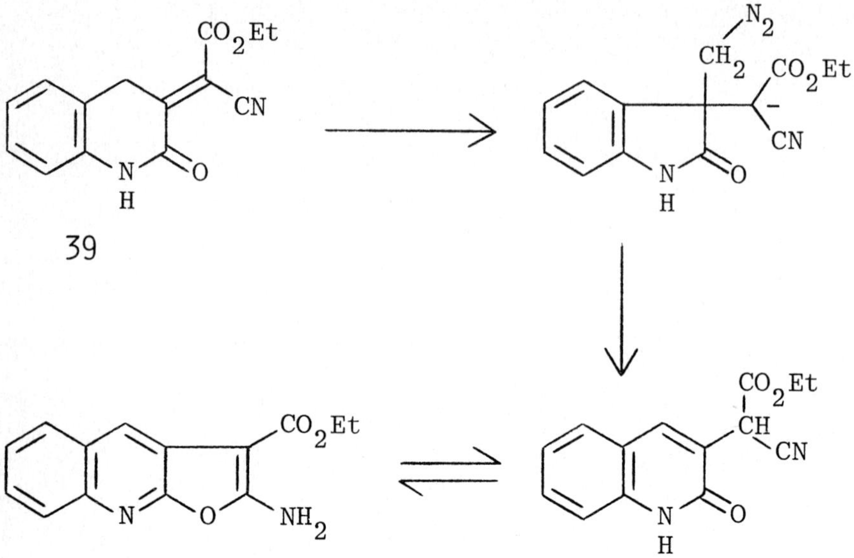

39

40

3,5,8-Trihydroxy-4-quinolone is the yellow zoochrome of the sponge Verongia aerophoba (G. Cimino, *et al.*, Tetrahedron Letters, 1984, 2925). This compound is readily air-oxidised to the blue quinone (41) which is highly unstable and polymerises to a black compound.

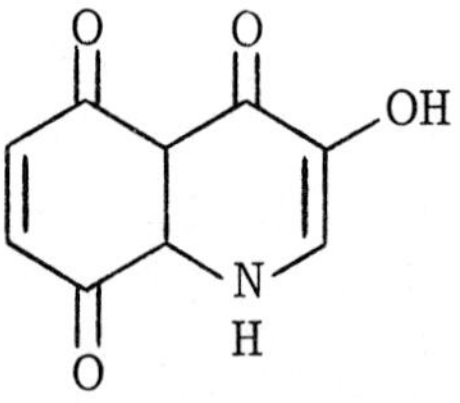

41

Other substituted 4-quinolones are the potential diuretic-
p-adrenergic blocking agents (42) for the synthesis of which
(43) is a key intermediate (A.K. Willand, R.L. Smith and E.J.
Cragoe, J. org. Chem., 1981, 46, 3846).

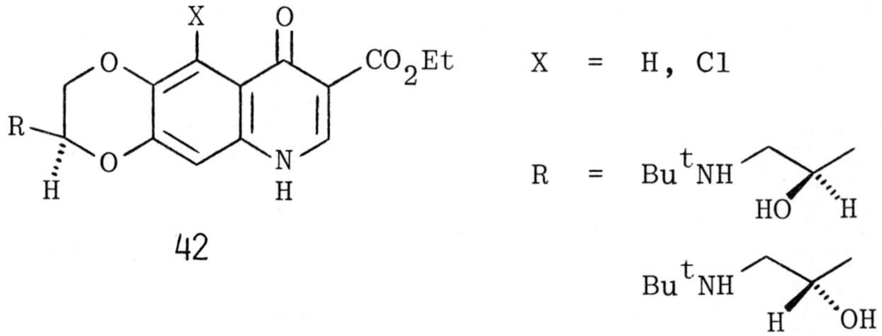

42

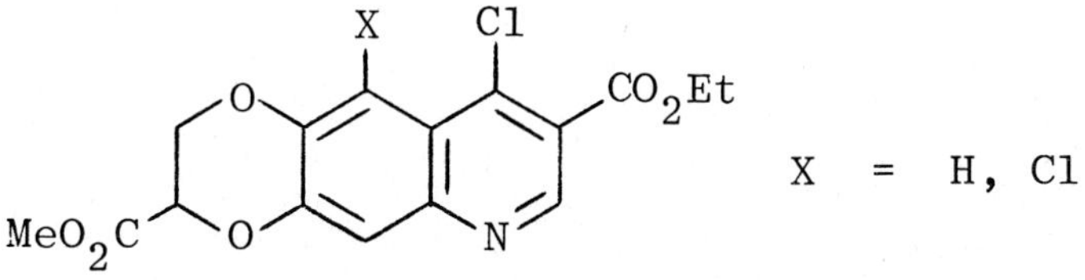

$X = H, Cl$

$R = Bu^tNH$ ... HO H

Bu^tNH ... H OH

43

$X = H, Cl$

5. Biquinolines

Many of the possible biquinolines were synthesised a number
of years ago (Dictionary of Organic Compounds, 5th Edition,
Chapman and Hall, New York, 1982, pp.684-5). Of the two
forms of 2,2'-biquinolines, *syn* (44) and *anti* (45) the
latter has been shown by X-ray crystallography, to be
preferred in the solid state (K. Folting and L.L. Merritt,
Acta Cryst., 1977, B33, 3540).

44   45

Phosphorescence studies by ESR spectroscopy of the triplet
state also reveal the predominance of the *anti-* form in
ethanol glasses at 77K (J. Higuchi, *et al.*, Bull. chem. Soc.
Japan, 1980, 53, 717). The 3,3'-annelated derivatives of
2,2'-biquinoline may be prepared by the reaction of -acetyl-
or *o*-benzoyl-aniline with cyclic α-diketones (P. Belser and
A. von Zelewsky, Helv., 1980, 63, 1675) yielding compounds
(46).

46   (n = 1, 2, 3, 4)

NMR spectroscopic studies (R.P. Thummel and F. Lefoulon, J.
org. Chem., 1985, 50, 666) show that compounds (46; n = 2-3)
undergo rapid conformational inversion at room temperature,
while (46; n = 4) is rigid. As the dihedral angle (α) between
the two quinoline rings becomes smaller H-8 becomes more
deshielded.

n	α°	δ (H-8, ppm)
1	0	8.46
2	20	8.43
3	50	8.35
4	65	8.25

Extrapolation of this relationship leads to dihedral angles of 130° and 120° for 2,2'-biquinoline and its 3,3'-dimethyl derivative respectively. The decrease in conjugative interaction in (46) as n is increased is reflected in the long wavelength UV-absorption bands of (46) which shift to shorter wavelength (higher energy) and become less intense.

Mono- and di-*N*-oxides of 2,2'-biquinoline and its 3,3'-annelated derivatives have also been prepared by Thummel and Lefoulon (*loc. cit*).

6. Hydrogenated quinolines

5,6,7,8-Tetrahydroquinolines are discussed in Chapter 25 as examples of cycloalkanopyridines.

The tetrahydroquinoline (47) is produced when the hydroxylamine (48) is treated with trifluoroacetic acid (M. Kawase and Y. Kikagawa, Chem. pharm. Bull. Japan, 1981, 29, 1615).

47

48

Tetrahydroquinoline itself can be electrophilically substituted in the 2-position *via* the formamidine derivative (49) (A.I. Meyers, S. Hellring and W.T. Hoeve, Tetrahedron Letters, 1981, 5115; A.I. Meyers and S. Hellring, *ibid.*, 1981, 5119; A.I. Meyers, Aldrichimica Acta, 1985, 18, 59).

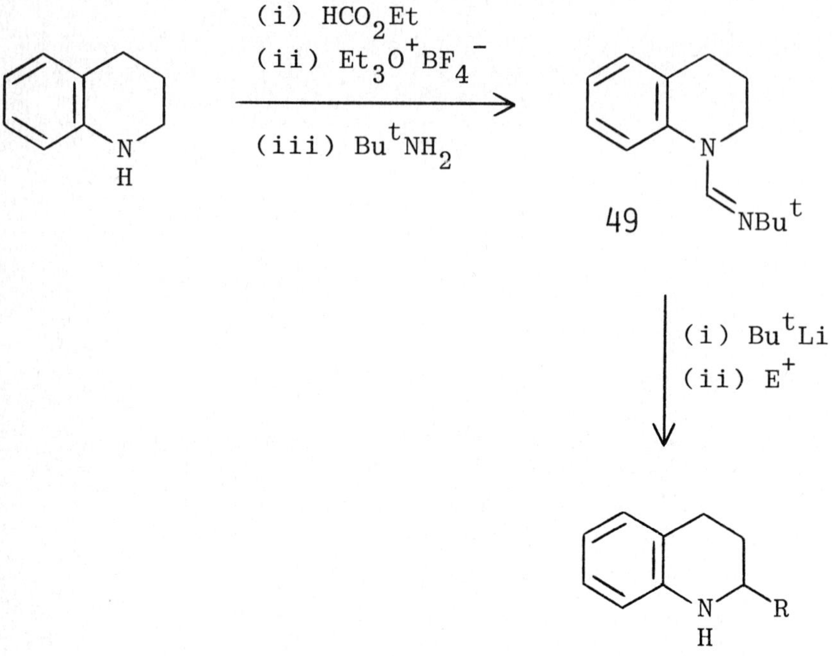

R = Me, Bu^n, PhCHOH

E^+ = MeI, Bu^nI, PhCHO

Anodic oxidation of N,N-dimethylaniline in methanol and subsequent treatment of the products with nucleophile/Lewis acid systems gives 4-substituted tetrahydroquinolines (T. Shono, J. Amer. chem. Soc., 1982, 104, 5753).

R' = H, R" = Ph; R' = Me, R" = Et; R' = $(CH_2)_3CH_3$, R" = $OSi(Me)_3$ etc.

A biomimetic course has been used for construction of the *cis*-decahydroquinoline ring system in synthesis of the poison-dart frog toxin pumiliotoxin C (M. Bonin, et al., Tetrahedron Letters, 1983, 1493).

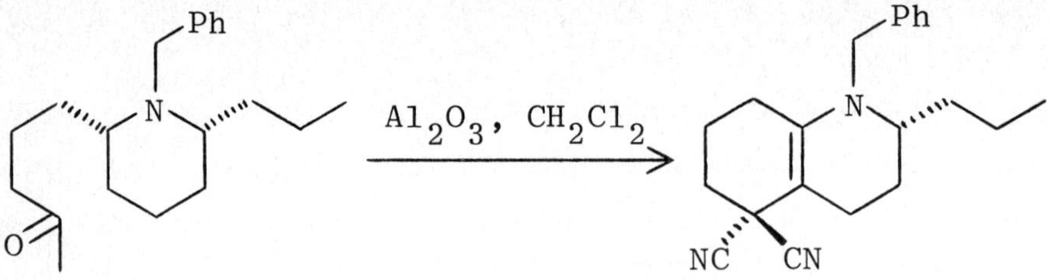

pumiliotoxin C

An elegant synthesis of *cis*-decahydroquinolines, shown below, also illustrates the stereo-specificity of the Diels-Alder reaction (D.L. Comins, A.H. Abdullah and R .K. Smith, *ibid*, p.2711).

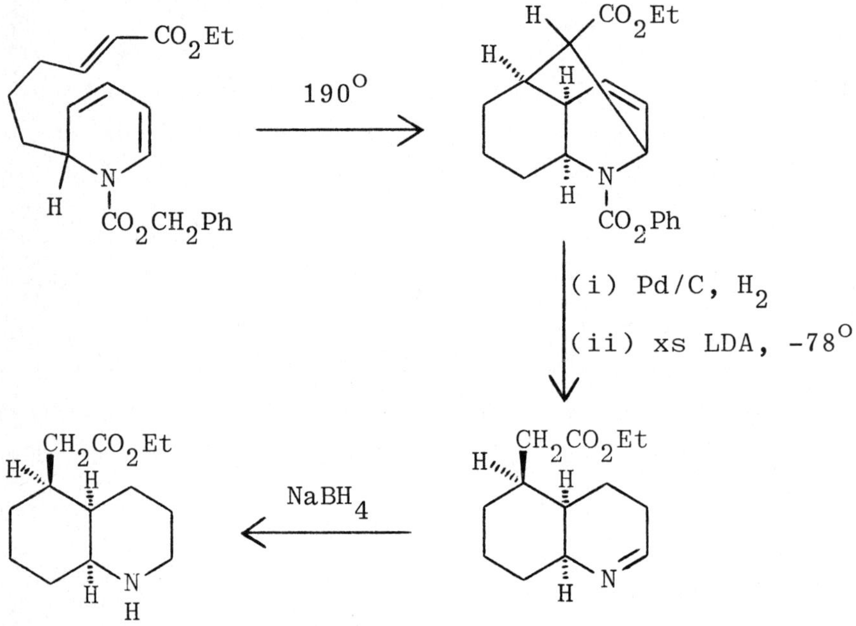

Typical examples for hydroquinoline formation incorporating various modes of reduction are given below.

Substrate	Reduction Method	Product	Ref.
Quinoline N-boro hydride	(i) $NaAl(OCH_2CH_2OCH_3)H_2$ (ii) $ClCO_2Me$	*N*-carbomethoxy-1,2-dihydro-quinoline	1,2
2-chloro-quinoline	Ca/MeOH	2-chloro-1,2,3,4-tetrahydro-quinoline	3
Quinoline	$C_5H_5\overset{+}{N}-BH_3$/HOAc	1,2,3,4-tetra-hydroquinoline	4
	$NaBH_4$/HOAC, AC_2O	*N*-acetyl-1,2-dihydroquinoline	5
	B_2H_6	1,2,3,4,-tetra-hydroquinoline	6
	transition metal (Mn Fe, CO) carbonyl hydride catalyts, CO, H_2O, base or CO, H_2	1,2,3,4-tetra-hydroquinoline	7

1. B.K. Blackburn, J.F. Frysinger and D. E. Minter,Tetrahedron Letters, 1984, 4913.
2. D.E. Minter and P.L. Stotter, J. org. Chem., 1981, 46, 3965.
3. H. Neunhoffer and G. Köhler, Tetrahedron Letters, 1978, 4879.
4. Y. Kikugawa, K. Saito and S.-I. Yamada, Synthesis, 1978, 447.
5. H. Katayama, M. Ohkoshi and M. Yasue, Chem. pharm. Bull. Japan, 1980, 28, 2226.

6. A. Nose and T. Kudo, Chem. Abs., 1980, 93, 26237).
7. R.H. Fish, A.D. Thormodsen and G.A. Cremner, J. Amer. chem. Soc., 1982, 104, 5234.

A number of studies (arbitrarily grouped here in Section 6 on hydrogenated quinolines, but equally eligible for Section 3(b)(ii) (nucleophilic substitution and addition) have been directed at quinoline analogues of the NAD$^+$ (nicotinamide adenine dinucleotide) NADH equilibrium.

1-Methyl or 1-benzylquinolinium compounds, bearing an electron-withdrawing group in the 3-position, give mixtures of the related 1,2- and 1,4-dihydroquinolines, the former predominating, when treated with NaBH$_4$, so that the reaction is a good preparative method for 1,2-dihydroquinolines (R.M.G. Roberts, D. Ostović and M.M. Kreevoy, J. org. Chem., 1983, 48, 2053). Alternatively, the 1,4-dihydroisomer can be formed by equilibrating the initial product mixture with a fresh sample of the quinolinium compound. 1,4-Dihydroquino-line derivatives have also been used to reduce the 10-methyl-acridinium ion (D. Ostović, et al., J. org. Chem., 1985, 50, 4206), and 2-pyridinecarbaldehyde to the corresponding alcohol (I. Tabushi, Y. Kuroda and T. Mizutani, J. Amer. chem. Soc., 1984, 106, 3377). The former reaction has been theoretically treated in terms of Marcus formalism for hydride transfer (M.M. Kreevoy and I.-S.H. Lee, J. Amer. chem. Soc., 1984, 106, 2550). Benzaldehyde can also be reduced to benzyl alcohol by 3-(N-α-methylbenzyl) carbamoyl-1,4-dimethyl-1,4-dihydroquinoline and its 2-methyl derivative in very high yield (A. Ono, et al., Tetrahedron Letters, 1982, 3185). An acid stable NADH model, 1-benzoyl-3-carbox-amido-1,4-dihydroquinoline, reduces benzylformic acid at low pH where the carboxyl group is undissociated (S. Shinkai, ibid., 1979, 3511); in turn, 3-substituted quinolinium salts may be selectively reduced to 1,4-dihydroquinolines using 1-benzyl-1,2-dihydroisonicotinamide (A. Nuvole, et al., J. chem. Res. (S), 1984, 356).

An analogous reaction is the regiospecific synthesis of quinolin-4-yl-phosphonates by treatment of N-(2,6-dimethyl-4-oxopyridin-1-yl)-quinolinium salts with tralkylphosphites in the presence of sodium iodide (A.R. Katritzky, J. chem. Soc., Perkin I, 1981, 668).

In a similar fashion, 1-ethoxycarbonyl-1,2-dihydro-quinoline-2-phosphonates may be regioselectively converted into 4-alkylquinolines (K.-y. Akiba, T. Kasai and M. Wada, Tetrahedron Letters, 1982, 23, 1709 - see Section 4(b)).

Chapter 27

Bicyclic compounds containing a pyridine; Isoquinoline and its derivatives

J. G. Keay and E. F. V. Scriven

1. *Isoquinolines*

This supplement covers major advances in isoquinoline chemistry that appeared in the literature during the period
1974 - 1985, thus completing a survey of the first hundred
years of isoquinoline chemistry. An excellent basic review
of the fundamentals of the chemistry of isoquinolines
appeared in 1979 (P. A. Claret in "Comprehensive Organic
Chemistry" Vol 4, ed. P. G. Sammes, Pergamon, Oxford, 1979,
p. 205). Another general review emphasizes industrial and
commercial aspects (S. N. Holter in "Kirk-Othmer Encycl.
Chem. Technol.", 3rd Ed. Vol. 19, eds. M. Grayson and D.
Eckroth, Wiley, New York, 1981, p. 532). Aspects of synthesis, structures, and reactivity of isoquinolines are covered
in "Comprehensive Heterocyclic Chemistry", Vol. 2, ed. A. R.
Katritzky and C. W. Rees, Pergamon, Oxford, 1984. A comprehensive treatment of isoquinoline chemistry has come out
in four parts in the "Chemistry of Heterocyclic Compounds"
series ("Isoquinolines", Vol. 38, Pt. 1-4, ed. G. Grethe,
Wiley, New York, 1981). More specialized reviews have
appeared on reactions of 1-substituted isoquinolines with
dimsylsodium (S. Kano, T. Yokomatsu, and S. Shibura, Heterocycles, 1977, 6, 1735); of isoquinolines with carbenes (U.K.
Pandit, Heterocycles, 1977, 6, 1520); and of anhydrides with
the carbon-nitrogen double bond (T. Shiraishi, T. Sakamoto,
and H. Yamanaka, Heterocycles, 1977, 6, 1716). The use of
isoquinoline derivatives as drugs has been described (A.
Brossi, Heterocycles, 1978, 11, 521). Isoquinoline alkaloids
are dealt with in Vol.IVH Chapter 36. However, several
reviews are of interest from the view point of isoquinoline
chemistry. They cover simple isoquinoline alkaloids (J.

184

Lundstroem, Alkaloids, 1983, <u>21</u>, 255), cyclizations to yield isoquinolines (W. Zielinski, Zesz. Nauk. Politech. Slask., Chem., 1981, <u>710</u>, 5), ^{13}C-nmr (D. W. Hughes and D. B. MacLean, Alkaloids, 1981, <u>18</u>, 217).

(a) Isoquinolines: Synthesis

Isoquinoline itself does not occur naturally but the inclusion of the isoquinoline moiety in alkaloids has spurred much work. The majority of methods reported for the preparation of iso-quinolines involves the cyclization of acyclic precursors to form the pyridine ring. Benzene derivatives are the common starting materials in many of these syntheses; and these form the classical methods of construction i.e. Bischler - Napieralski, Pictet - Gams reaction etc.

Synthetic aproaches to isoquinolines have been reviewed (T. J. Kametani and K. Fukumoto, in "The Chemistry of Heterocyclic Compounds, Isoquinolines", ed. G. Grethe, John Wiley & Sons, New York, Vol. 38, Part 1, p. 139-274). Systematically isoquinoline syntheses can be divided into two types involving the formation of each ring. This may be achieved by single or multiple C-C or C-N bond formation; cycloaddititon reactions or by rearrangement.

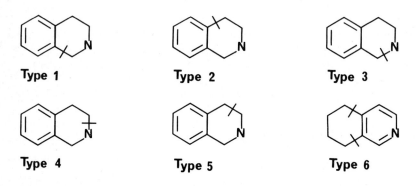

Type 1   Type 2   Type 3

Type 4   Type 5   Type 6

Type 7; **Rearrangements**

(i) Type 1 Syntheses

(1) The Bischler-Napieralski reaction.

This remains the most frequently employed method for the
synthesis of isoquinolines.   The reaction involves the
cyclodehydration of *N*-acyl-2-phenethylamines with a Lewis
acid in a solvent (A. Bischler and B. Napieralski, Ber.,
1893, 26, 1903). The product is a 3,4-dihydroisoquinoline.

As originally described the Bischler-Napieralski reaction is
carried out at high temperatures with phosphorus (V) oxide
or chloride. Moderation of the conditions and the use of
better dehydrating agents has resulted in dramatic yield
increases.   Aromatic hydrocarbons, nitrobenzene and haloalkanes have been used as solvents.   Phosphorus oxychloride has been used without a solvent. Acetonitrile has
also been employed giving high yields (T. Kametani and co-
workers, J. chem. Soc. (C), 1971, 3350). Lewis acids which
have been used include PCl_5, $POCl_3$, polyphosphoric acid
(PPA) and its ester (PPE), $AlCl_3$, $SnCl_4$ and even HCl.

3,5-Dimethoxybenzylalkyl ketones on treatment with acetamide give 3-alkyl-1-methylisoquinolines (T. Hirota *et al.*,
Chem. Pharm. Bull., 1978, 26, 245). The appeal of this procedure is that the substituents at two positions (i.e. 1- and
3-) can be varied from simple starting materials.

The cyclodesulphurization of thioamides has been employed as
a mild alternative in the Bischler-Napieralski reaction.
N-(1-Methyl-2-(3,4-dimethoxyphenyl)ethyl)thiobenzamide gives

the corresponding 3,4-dihydroisoquinoline derivative on treatment with mercury (II) chloride or phosphorus oxychloride (M.S. Ragab *et al.*, Pharmazie, 1974, 29, 178).

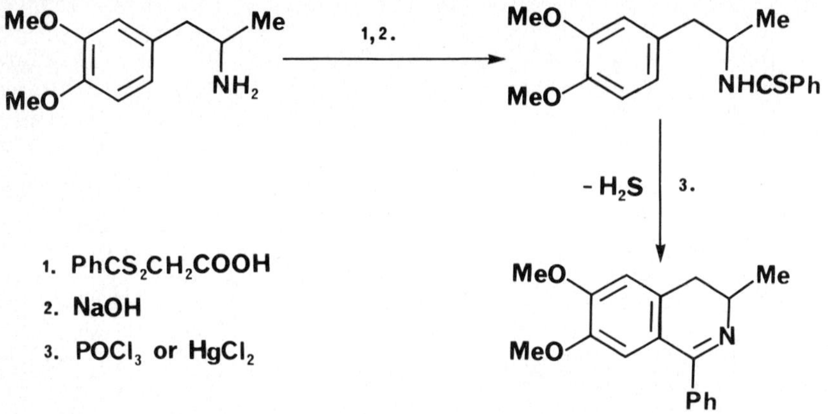

1. PhCS$_2$CH$_2$COOH
2. NaOH
3. POCl$_3$ or HgCl$_2$

Beckmann rearrangement of the oxime shown below and cyclization of the intermediate imidoyl chloride with phosphorus (V) oxide gives a one-pot preparation of 1,3-dimethylisoquinolines (W. Zielinski, Synthesis, 1980, 70).

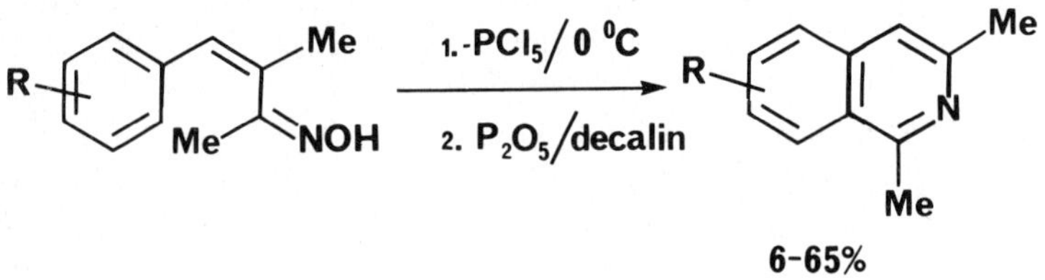

6-65%

In addition to 3,4-dihydroisoquinolines, *N*-(2-chlorophenyl) ethyl amides are also obtained on treatment of the *N*-2-phenylethyl amides with phosphorus (V) chloride at reflux in toluene (W. Wiegrebe, *et al.*, Helv., 1975, 58, 1825). Using model compounds this behaviour is not observed with phosphorus oxychloride, but only occurs when excess phosphorus pentachloride is employed after 48h at reflux in toluene (A. Berger, G. Dannhardt, and W. Wiegrebe, Arch. Pharm., 1984, 317, 488).

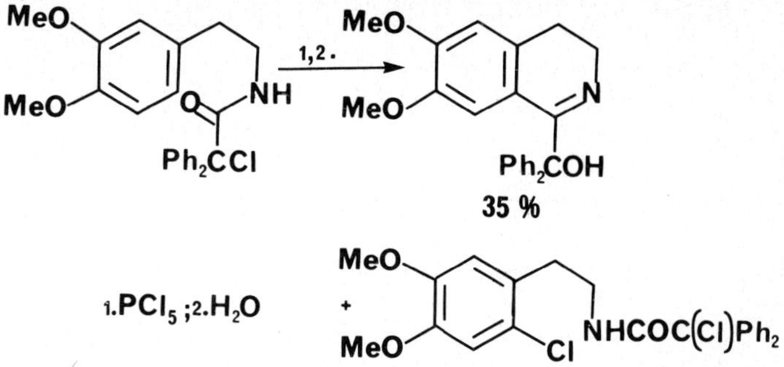

Carbamates and isocyanates have been used with phosphorus oxychloride and tin (IV) chloride to give lactams (Y. Tsuda *et al.*, Heterocycles, 1976, 5,157; K. Isobe, J. Taga, and Y. Tsuda, *ibid.*, 1978, 9, 625). This combination of reagents gives yields between 60-90% at temperatures of 100-150°C. Higher temperatures afford halogenated and fully aromatic by-products.

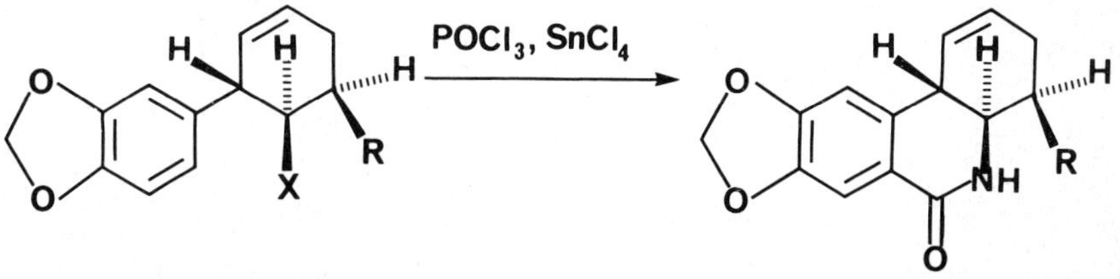

X = NCO, NHCO₂Me

R = H, COOMe, CH₂COOMe

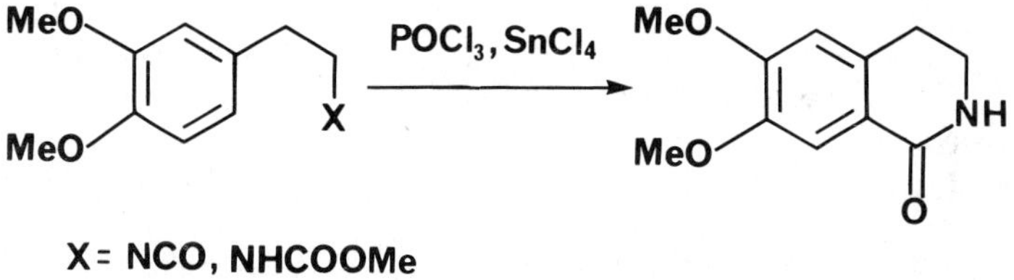

X = NCO, NHCOOMe

Unsaturated carbamate esters (*e.g.* allyl carbamates) can be converted into fused tetrahydroisoquinolines. These reactions are believed to proceed *via* *N*-acyliminium salt intermediates (S. Kano, Y. Yuasa, and S. Shibuya, Synth. Comm., 1985, 15,883).

1. $O_3, CH_2Cl_2, -78\ ^0C$; **2.** Me_2S; **3.** HCOOH.

Treatment of the carbamate derivative (1) with glyoxylic acid followed by hydrogenation gives the tetrahydroisoquinoline (2) in 80% yield (S. Danishefsky *et al.*, Tetrahedron Letters, 1980, <u>21</u>,4819).

1. HCOOH, **2.** $Cl_2CHCOOH$, **3.** CH_2N_2, **4.** H_2/Pd-C.

Isonitriles (4), which can be prepared by the reaction of dichlorocarbene on 2-phenethylamines, produce 3,4-dihydro-isoquinolines in good yield when cyclized with certain Lewis acids (Y. Ban, T. Wakamatsu, and M. Mori, Heterocycles, 1977, <u>6</u>, 1711).

	Yield(%)
$R^1 = R^2 = CH_2$,	70
$R^1, R^2 = CH_3$,	30

Isonitriles react with acyl halides under mild conditions in the presence of a silver salt to give, *via* the ketoimidoyl halide, good overall yields of 1-acyl-3,4-dihydroisoquinolines (M. Westling and T. Livinghouse, Tetrahedron Letters, 1985, 26,5389).

$$1. CH_2Cl_2, 0-25\ ^0C, 0.5-18h;\ 2.\ Ag^{\oplus}CF_3SO_3^{\ominus}, 12h.$$

The importance of silver trifluoromethanesulphonate has been recorded in the reaction of allylbenzenes with nitriles (T. Sako, K. Tamura, and K. Nagayoshi, Chem. Letters, 1983, 791). The triflate anion is a poor nucleophile and its limited association with the intermediate nitrilium salt offers little hindrance to cyclization and formation of the 3,4-dihydroisoquinoline.

1. $AgOTf-I_2/CH_2Cl_2$; 2. R^2CN; 3. $KOH/MeOH$

$R = H, Me$; $R^1 = H, 4-MeO$; $R^2 = Me, Et, Ph$.

Other variants of the Ritter reaction have been applied to isoquinoline synthesis. These proceed both thermally and photochemically (at lower temperatures) but the highest conversions are achieved under a combination of conditions (T. Kitamura, S. Kobayashi, and H. Taniguchi, Chem. Letters, 1984, 1351).

R = H, MeO; R¹ = Me, Et, Ph.   52 - 96 %

Bischler-Napieralski cyclization of *N*-(2-(2-naphthyl)ethyl) amides (5) can clearly give two different products the linear dihydrobenz*[g]*-(6) or the angular dihydrobenz*[h]*-isoquinoline (7). In these cases only the angular derivatiyes (7) were obtained despite the use of blocking groups (R^2=Br,SEt) which were expelled prior to cyclization (R.D. Waigh and D. Beaumont, J. chem. Res. (S), 1979, 332).

R^1= H, CH₂Ph, CHPh₂, OEt, Ph.

R^2= H, Br, SEt.

R^3, R^4, R^5= H, OMe.

Attempted cyclization of the *N*-(3-acetoxy-4-benzyloxy phenethyl)phenylpropionamide (8) to the 3,4-dihydroiso-quinoline (9) gives the indane derivative (10) (T. Kametani, Y. Satoh, and K. Fukumoto, Chem. Pharm. Bull., 1977,25,1129). Competitive ring closure by the more electron rich moiety results in the more favoured ring closure.

Cyclization of a 1,2-diarylethylamide to a 3-aryl-3,4-dihydro-isoquinoline is usually unsuccessful unless the formamide derivative is used or that phenyl ring which is to become the 3-substituent of the isoquinoline contains no substi-tuents. Generally, stilbenes and stilbene dimers are generated. However, if the reaction is carried out at 0°C with phosphorus (V) chloride using a nitrile as solvent then the dihydroisoquinoline(11) is formed. Interestingly, the major product is that which involves incorporation of the solvent, and is formed in 40-53% with the minor product (12) in 7-9%. This lends support to the intermediacy of a nitri-lium salt.

1. PhMe – P_2O_5 or $POCl_3$ or PCl_5;
2. R^1CN, PCl_5 – 0 °C.

R = Me, Ph, PhCH$_2$
R^1 = Me, Et, Ph, PhCH$_2$

Isoquinolin-3-ols are produced when phenylacetyl chlorides are treated with alkyl or arylthiocyanates in the presence of tin (IV) chloride. The preferred solvent is nitrobenzene and yields up to 65% have been achieved (M.A. Ainscough and A.F. Temple, Chem. Comm., 1976, 695). With cyanogen chloride the only isolable product is 3-amino-1-chloroisoquinoline in 38% yield.

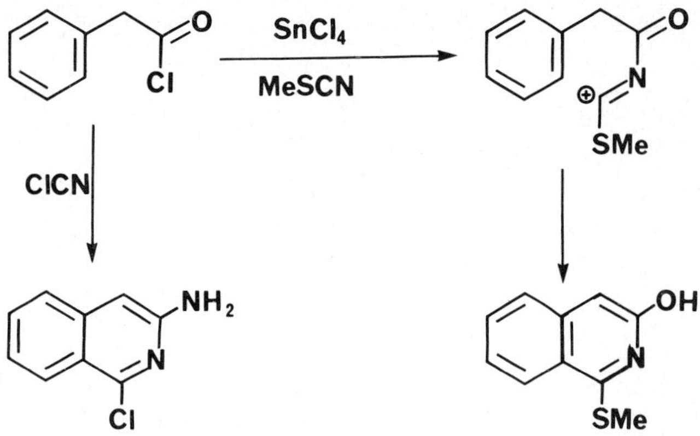

i) Mechanism

The isolation of nitriles and haloalkanes from Bischler-
Napieralkski reaction mixtures has allowed the reaction
mechanism to be elucidated and allowed this and related
reactions to be unified through a common intermediate. In
the synthesis of 1-(3,4-dimethoxybenzyl)-3,7-dimethyl-
5,8-dimethoxy-3,4-dihydroisoquinoline; 3,4-dimethoxyphenyl
acetonitrile(13) and 1-(2,5-dimethoxy-4-methylphenyl)-
2-chloropropane(14) are formed (J. Gal, R. J. Weinkam, and
N. Castagnoli, Jr., J. org. Chem., 1974, 39,418). Such products arise from the von Braun amide degradation and their
formation suggests that a common intermediate (15) may exist
for both processes.

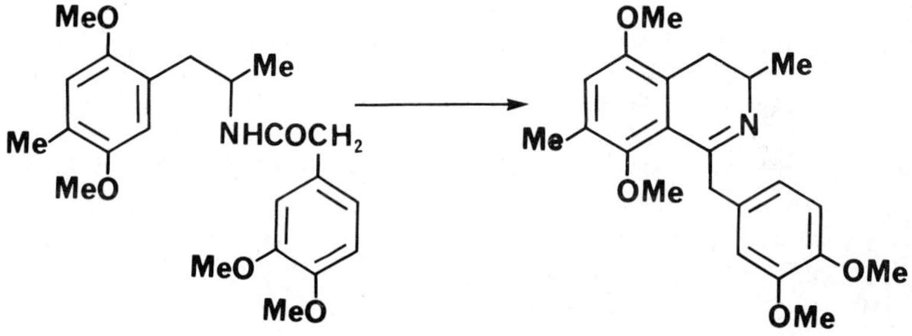

$+ (13),(14),(15)$

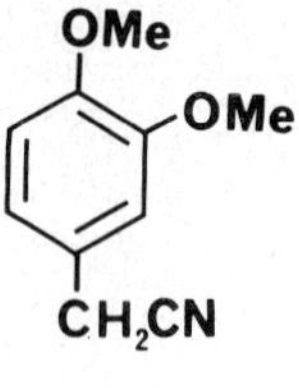

(13) 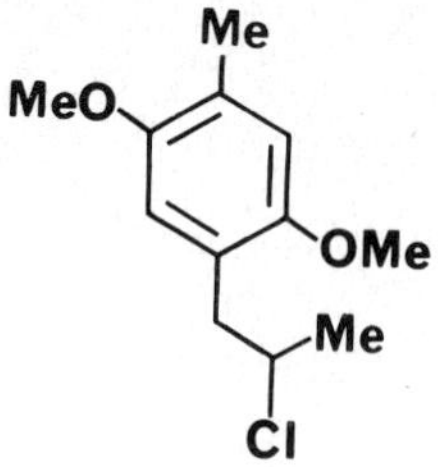　(14) 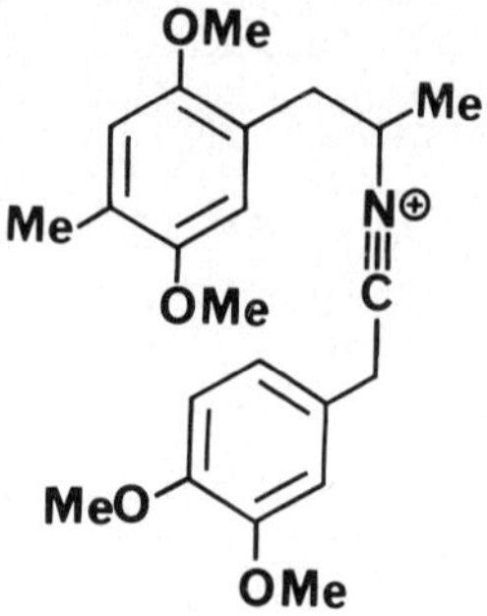　(15)

Nmr evidence has been obtained to show the presence of a nitrilium ion in certain Bischler-Napieralski reactions (G. Fodor, J. Gal, and B. A. Phillips, Angew. Chem. intern. Edn., 1972, 11,919; P. J. Stang, and A. G. Anderson, J. Amer. chem. Soc., 1978, 100,1520). The reaction is now believed to proceed by formation of nitrilium salt intermediates. The amide is first converted into an imidoyl halide which then forms a nitrilium ion when treated with a Lewis acid (G. Fodor and S. Nagubandi, Tetrahedron, 1980, 36,1279).

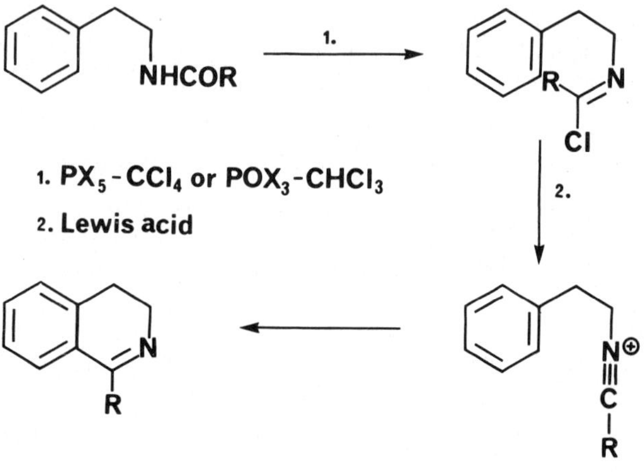

1. $PX_5 - CCl_4$ or $POX_3 - CHCl_3$

2. Lewis acid

These same authors have used this approach to correlate a series of heretofore unrelated reactions.

$X=Cl$, $OCOCF_3$, OSO_2CF_3, OPO_2PO_3H; $R=Aryl$

a. Lewis acid
b. PCl_5, $POCl_3$, $SOCl_2$, $(CF_3CO)_2O$, CF_3SO_2Cl, PPE
c. von Braun reaction
d. Retro-Ritter reaction
e. Ritter reaction
f. Beckmann reaction
g. Bischler-Napieralski reaction
h. Schmidt reaction

Thus, the von Braun amide degradation can be seen as occurring *via* degradation of the same intermediate nitrilium salt and explains the presence of nitriles and alkyl halides commonly observed as products in the Bischler-Napieralski reaction. Similarly, the anion can act as a base, generating the olefin instead of the alkyl halide in a retro-Ritter reaction. When R=phenyl such behaviour is favoured due to the fully conjugated *trans*-stilbene generated (N. S. Narasimhan, M. S. Wadia, and N. R. Shete, Ind. J. Chem., 1980, 19B, 556).

Equilibration of the α-chloroimines generated from *N*-(2-(3,4-dimethoxyphenyl)ethyl)-2,2-dimethylpropionamide and phosphorus (V) chloride has shown that the Z-(anti) form is the preferred configuration (U. Berger, G. Dannhardt, and W. Wiegrebe, Arch. Pharm., 1983, 316, 182; *ibid.*, 1984, 317, 488). Treatment of the same amide with phosphorus oxychloride gives an isolable intermediate (16). All these intermediates give the dihydroisoquinoline on heating in carbon tetrachloride.

Z (anti) (16)

New reagents have been introduced during this study including polyphosphoric ester, trifluoroacetic acid anhydride and trifluoromethylsulphonic anhydride (S. Naqubandi and G. Fodor, Heterocycles, 1981,15,165). For most cases the convenience of a one-pot procedure is sacrificed for a two-step procedure which gives higher yields under much milder conditions. The intermediate imidoyl halide prepared from amide and phosphorus (V) chloride is treated with an excess (*ca* 15 eq.) of the Lewis acid (*e.g.* $SnCl_4$) in anhydrous chloroform.

ii) Reagents

Choice of reagents is determined by the structures' functionalities. Typically Bischler-Napieralski reactions are carried out on the amide with phosphorus (V) oxide, phosphorus oxychloride or polyphosphoric acid in xylene or toluene at reflux temperatures. However, milder conditions cause less product decomposition and fewer side reactions. Under the usual reaction conditions demethylation of phenolic ethers may occur with polyphosphoric acid (N. H. Martin and C. W. Jefford, Helv., 1982, 65, 762).

iii) Orientation

The orientation of the Bischler-Napieralski reaction is that the closure is facilitated by strongly electron donating groups (e.g. HO, RO, SR, NHR) ortho and para to the site of closure. Preference is for the position para to the electron donating group.

iv) Stereochemistry

Stereochemical control during the Bischler-Napieralski reaction has received little attention. However, some work has been carried out using L-N-acetyl-3,4-dimethoxyphenyl-alanine methyl ester (17). In terms of yield the best cyclization procedure involves the use of phosphorus (V) oxide in refluxing xylene. However, the product shows no optical rotation and racemization does not occur merely in refluxing xylene. Partial racemization occurs when polyphosphate ester is used as a cyclizing agent. Treatment with phosphorus oxychloride in acetonitrile for 2h at room temperature gives the dihydroisoquinoline without racemization (T. Kametani et al., Heterocycles, 1982, 19, 535).

(2) The Pictet-Gams reaction

This modification of the Bischler-Napieralski reaction involves the cyclodehydration of 2-hydroxy-2-phenethylamides. In this case fully aromatic isoquinolines might be the expected product.

However, this reaction has now been shown to proceed *via* a Δ^2-oxazoline which only rearranges to the isoquinoline under the more forcing conditions of reflux with phosphorus (V) oxide in decalin (N. Ardabilchi *et al.*, J. chem. Res. (S), 1982,156). As previously noticed isomeric isoquinolines are also isolated from the reaction mixture. The migration occurs during the oxazoline to isoquinoline rearrangement and it is proposed that this proceeds *via* a vinylamide which in some cases may be isolated. Earlier work carried out by the original authors (W. M. Whaley and W. H. Hartung, J. org. Chem., 1949,14,650) involving the cyclization of *erythro*-2-benzamido-1-phenylpropan-1-ol has been repeated and corrected (N. Ardabilchi *et al.*, J. chem. Soc., Perkin I, 1979,539).

ArCH=C(R)NHCOPh   ArC(R)=CHNHCOPh

The migratory aptitude is related to the electron releasing capacity of the group itself. An amide possessing an electron releasing group in the 2-position is likely to undergo extensive rearrangement. Similarly, cylizations involving 2-benzamido-1-methoxy-1-phenylalkanes involve the formation of the Δ^2-oxazolines.

(3) The Pictet-Spengler reaction

The condensation of a 2-phenethylamine with an aldehyde or ketone in strong acid leads to the formation of a 1,2,3,4-tetrahydroisoquinoline. This is the standard method for the synthesis of tetrahydroisoquinolines, and many derivatives have been prepared by this technique (M. Shamma and J. L. Moniot, "Isoquinoline Alkaloid Research: 1972 - 1977"; Plenum Press: New York, 1978; M. Shamma, Org. Chem. (N.Y.), 1972, 25; T. Kametani, "The Total Synthesis of Natural Products", ed. J. W. ApSimon, Wiley, New York, 1977).

Activation by alkylthio substituents has been found to improve the yield of isoquinolines on cyclization. These groups may also be removed much more readily that their oxygen counterparts. The site selectivity between methylthio and methoxy groups has also been compared (M. R. Euerby et al. , J. chem. Soc., Perkin II, 1985, 1151).

Dopamine forms isosalsolinol when treated with an excess of acetaldehyde at controlled pH (G. S. King, B. L. Goodwin, and M. Sandler, J. Pharm. Pharmacol., 1974, 26, 476).

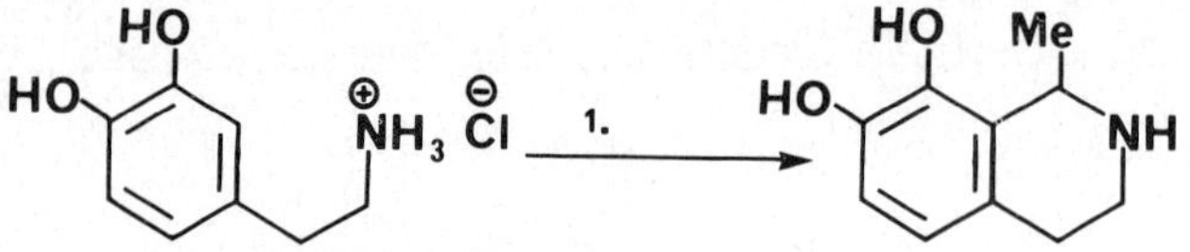

1. **MeCHO, pH 4·5, 20 °C.**

As an alternative to Kametani's procedure for the preparation of a 1,1-disubstituted-1,2,3,4-tetrahydroisoquinoline, homo-veratrylamine is treated with a dialkyl acetylenedicarboxylate and the adduct cyclized with polyphosphoric or sulphuric acid to give a 6,7-dimethoxyisoquinoline. This reaction may be limited to activated substrates (M.D. Nair and J. A. Desi, Ind. J. Chem., 1979, 17B, 277).

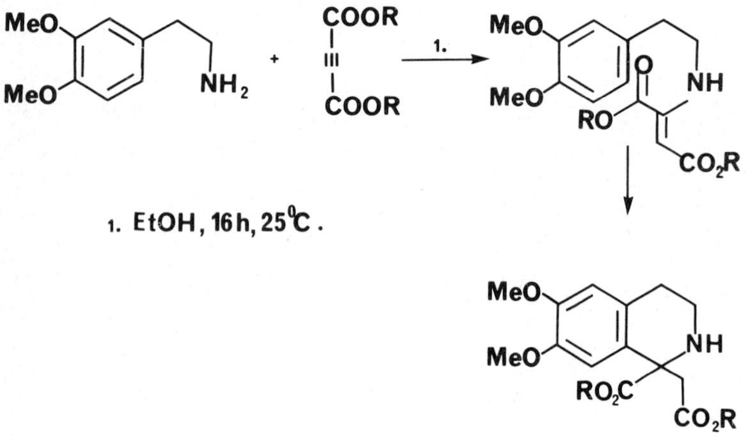

1. **EtOH, 16 h, 25 °C .**

7-Hydroxy-6-methoxy-2-methyl-1,2,3,4-tetrahydroisoquinoline (corypalline) has been prepared *via* a Pictet-Spengler reaction (L. J. Fransisca Mary *et al.*, Ind. J. Chem., 1977, 15B, 182). Tetrahydroisoquinolines are available by similar procedures (S. Ruchirawat *et al.*, Synth. Comm., 1984, 14, 1221; I. W. Mathison, W.E. Solomons, and R.H. Jones, J. org. Chem., 1974, 39, 2852).

1. $HC(OH)_3$, $HCOOH$; 2. $(COOH)_2$.

Due to the interesting physiological effect of many phenethyl-
and tryptylamine derived isoquinoline and 2-carboline products,
attempts have been made to effect the Pictet-Spengler type
cyclization under moderate conditions. 4-Hydroxytetrahydro-
isoquinolines are a good example of where advances have been
made; the labile benzylic hydroxyl group leads to low yields
and numerous by products. Normal Pictet-Spengler conditions
with Schiff's base and aldehydes in MeCN, MeOH, AcOH and
DMF give complex product mixtures. However, ethanol containing a catalytic amount (1 %) of silica gel yields in
excess of 80% of the isoquinoline after 2h at 90°C. Without
silica gel approximately 8h is required to achieve the same
conversion. Preforming the Schiff's base is unnecessary (T.
Hudlicky *et al.*, J. org. Chem., 1981,**46**, 1738).

1. **Silica gel, EtOH, 90°C, 2h; R=OH, H; Yield = 87, 92%**

Tetrahydroisoquinoline derivatives containing a 3-carboxy and a 4-hydroxy functionality are prepared from piperonal and glycine. The intermediate diastereoisomers, the *threo* isomer predominates (4:1), can be separated prior to cyclization (R. Sola *et al.*, Heterocycles, 1982, <u>19</u>, 1797).   Nmr-coupling constants can be used to distinguish between the isomers.

1. **NaOH, EtOH, 25 °C, 8h;**
2. **aq. HCHO, 1N H_2SO_4, 25 C.**

Tetrahydroisoquinoline 1-carboxylic acids may be oxidatively decarboxylated electrochemically to yield 3,4-dihydro-isoquinolines. These may then be chemically reduced to the tetrahydroisoquinolines. Oxidations are carried out in 0.1 M NaOMe/MeOH or $NaHCO_3/H_2O$-MeOH; yields are typically 50-90%. A direct correlation between the ease of oxidation and the number of free hydroxy groups on the aromatic ring can be made (J. M. Bobbitt, and T. Y. Cheng, J. org. Chem., 1976, <u>41</u>, 443).

The Pictet-Spengler reaction has been applied to carbohydrates with sugars being successfully reacted with 2-phenethylamines and tryptamines (D. B. MacLean, W. A. Szarek, and I. Kvarnstrom, Chem. Comm., 1983, 601; F. Czarnocki, D. B. MacLean, and W. A. Szarek, Chem. Comm., 1985, 1318). Dopamine hydrochloride and D-glucose heated at reflux in water give after chromato-graphy a 92% yield of a mixture of diastereoisomers (4:1). 2,5-Anhydro-D-mannose yields (18) in 98% yield. The reaction of catecholamines with aldehydes is believed to occur readily under physiological conditions and may in part contribute to the behaviour-related symptoms of ethanol abuse (H. A. Bates, J. org. Chem., 1983, 48, 1932).

1. D-glucose, H₂O, 48h.

(18)

Norepinephrine reacts with both formaldehyde and acetaldehyde. At pH 0.7 formaldehyde gives only one isomer (19a) but at pH 6-7 the isomeric tetrahydroisoquinoline (20a) is also formed in a 1:4 ratio. The rate of reaction is much slower at the lower pH. With excess acetaldehyde at pH 6.5 a 5:2 mix of products is obtained.

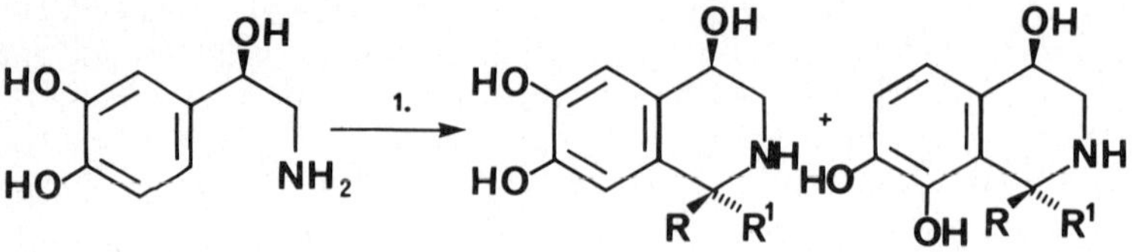

1. RR^{1}CO, H$_2$O.

(19)

(20)

R,R^1 = H,Me.

	a	b	c
R	H	Me	H
R^1	H	H	Me

Oxazolidines and thiazolidines have been employed as "one carbon fragments" yielding isoquinolines on reaction with 2-phenethylamines (H. Singh and R. Sarin, Heterocycles, 1985, 23, 107).

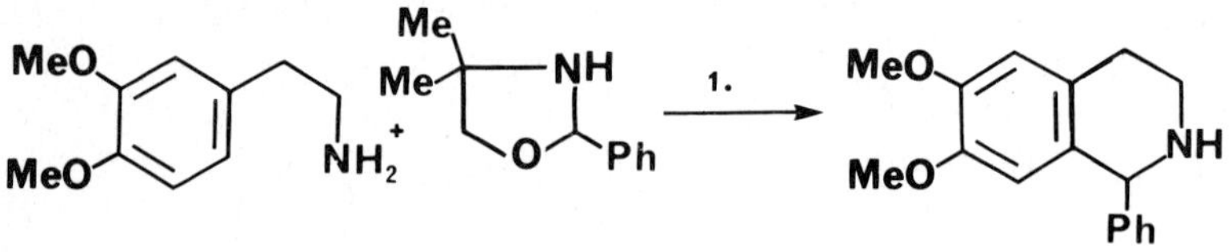

1. CF$_3$COOH, MeCN, Δ/10h.

(ii) Type 2 Syntheses

(1) The Pomeranz-Fritsch reaction

The original two step procedure developed by Pomeranz and Fritsch (C. Pomeranz, Monatsh. 1893, 14, 116; P. Fritsch, Ber., 1893, 26, 419) provides a general synthesis for isoquinolines. An aromatic aldehyde is condensed with an aminoacetaldehyde acetal and the product cyclized with mineral acid.

Bobbitt's modification, *via* reductive cyclization although longer, proceeds *via* mild cyclization of benzylamine acetals and *in situ* reduction to tetrahydroisoquinolines (J.M. Bobbitt *et al.*, J. org. Chem., 1965, 30, 2247). Dehydrogenation to the isoquinoline is often not simply achieved. Prior to this, other workers had employed a two-step process from benzylamines and glyoxal acetals (E. Schlitter and J. Müller, Helv., 1948,31,914,1119). Employing the observation of Jackson that *N*-tosyldihydroisoquinolines readily eliminate *p*-toluenesulphonic acid (A. J. Birch, A. H. Jackson, and P.V.R. Shannon, J. chem. Soc., Perkin I, 1974, 2185,2190), Boger has developed a two-step isoquinoline synthesis from benzyl halides (or mesylates) and *N*-tosyl aminoacetaldehyde dimethyl acetal (D. L. Boger, C. E. Brotherton, and M. D. Kelley, Tetrahedron, 1981,37,3977).

1. **NaH - DMF, 25°C.**

2. **Dioxane 6N HCl, Δ/24h.**

X = halogen, OMs

58 - 79%

Yields in Pomeranz-Fritsch type reactions depends upon the nature of the substituent carried by the benzene ring. With weakly donating groups (alkyl, halogen) competitive oxazole formation may occur. The presence of nitro groups leads to exclusive oxazole formation (E. V. Brown, J. org. Chem., 1977, 42,3208).

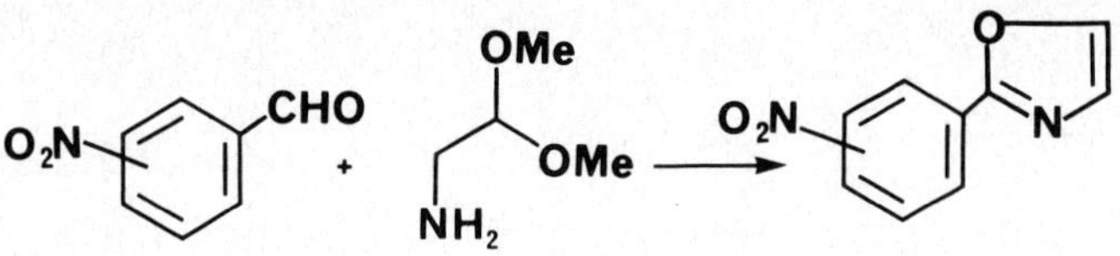

The cyclization of *N*-benzyl diethoxyacetamides in sulphuric acid provides a high yielding route to 3-isoquinolinones (H. Fukumi and H. Kurihara, Heterocycles, 1978, 9, 1197).

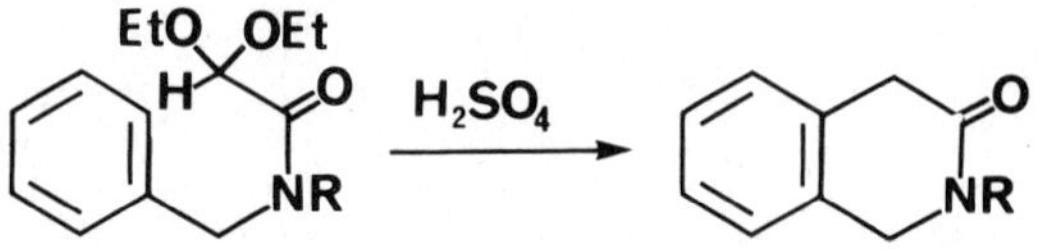

For synthetic convenience one-pot procedures have been employed. The intermediate imines are sequentially treated with ethyl chloroformate, trimethyl phosphite and the car-bamate phosphonates cyclized with excess titanium (IV) chloride (J. B. Hendrickson and C. Rodriquez, J. org. Chem., 1983, 48, 3344).

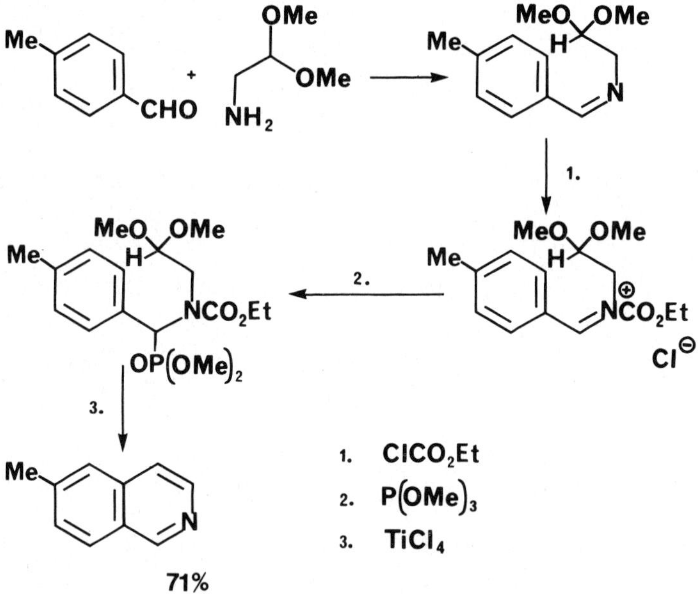

1. $ClCO_2Et$
2. $P(OMe)_3$
3. $TiCl_4$

71%

(2) Miscellaneous Methods

i) From organometallic reagents

Organopalladium complexes have been employed for intramolecular carbon-carbon bond formation. Treatment of the
2-bromoaniline (21) with palladium acetate and triphe-
nylphosphine in the presence of TMEDA and in an inert
atmosphere gives the isoquinoline in 27% yield. The benzylidene derivative is also obtained (8%) (M. Mori, K.
Chiba, and Y. Ban, Tetrahedron Letters, 1977,1037; M. Mori
and Y. Ban, Tetrahedron Letters, 1979, 1133).

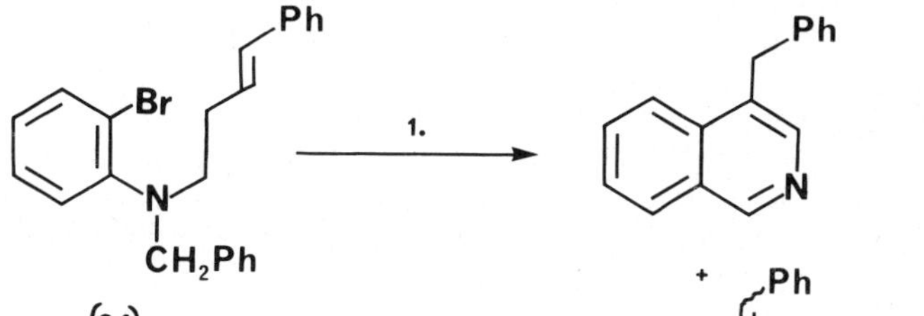

1. Pd(OAc)$_2$,Ph$_3$P,TMEDA,
125°C, 96h.

ii) Photolysis of halobenzylamines

Photocyclization of the carbanion generated from an
N-acetyl-*N*-methyl-*o*-chlorobenzylamine gives a 1,4-dihydro-
3-(2*H*)isoquinoline in 54-62% yield (R. R. Goehring *et al.*,
J. Amer. chem. Soc., 1985,107,435).

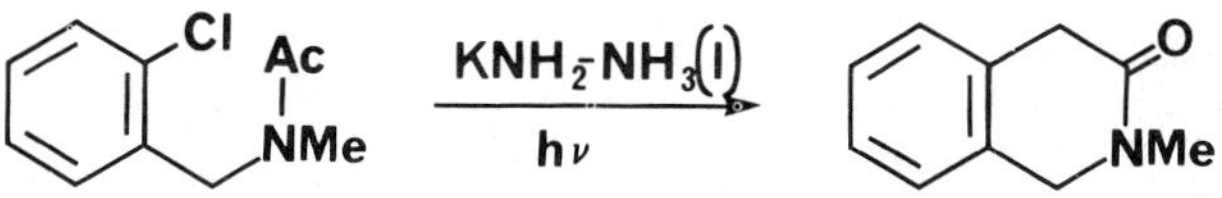

KNH$_2$,NH$_3$(l)
hν

iii) Via carbamate cyclization

The treatment of the carbamate (22) with methanesulphonic acid
generates the 3-isoquinolinone in high yield (D. Ben-Ishai, N.
Peled, and I. Sataty, Tetrahedron Letters, 1980,21,569).

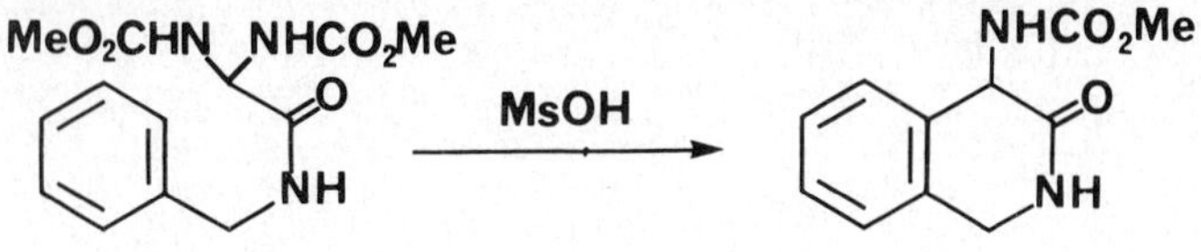

(22)   84%

iv) Cyclization of N-(3-bromopropyl)benzylamines

Aluminium chloride reacts with 1-bromo-3-(4-chlorobenzyl-amino)propane leading to a mixture of the isoquinoline and the benzazepine (64% and 14% respectively) (C. D. Perchonock and J. A. Finkelstein, J. org. Chem., 1980,45,2000).

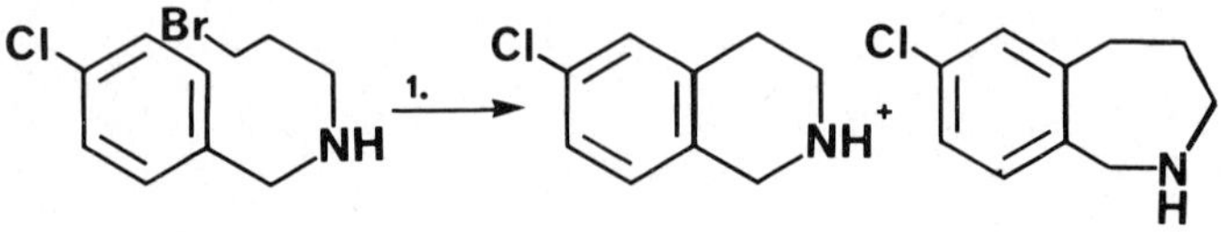

1. AlCl$_3$.

v) Flow vacuum thermolysis of 2-azabuta-1,3-dienes

An annelation procedure has been described in which 2-aza-1,3-butadienes are pyrolyzed at 600°C to the 3,4-dihydro-isoquinolines in moderate yields (C. K. Govindan and G. Taylor, J. org. Chem., 1983,48,5348). Thermolysis of phenyl-2H-azirines produces similar products which are believed to derive from intermediate 2-azabutadienes (R. G. Bergman and L. A. Wendling, J. org. Chem., 1976,41,831). The presence of electron-withdrawing groups on the benzene ring usually prevents ring closure.

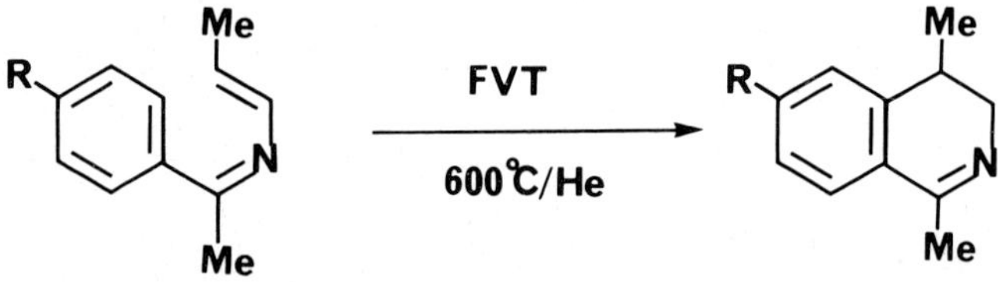

45 - 65%

R = H, OMe, Me, F, Cl, CN.

vi) Via nitrosation of indanones

3-Substituted indan-1-ones on treatment with butyl nitrite in the presence of sodium methoxide undergo a ring enlargement to the isocarbostyrils (J. N. Chatterjea *et al.*, Ann., 1981,52).

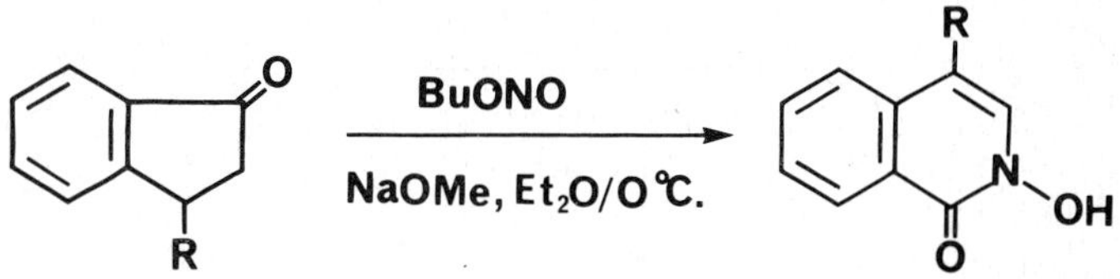

O
R
BuONO
NaOMe, Et$_2$O/0 ℃.
R
N
OH
O

(iii) Type 3 Syntheses

(1) The Gabriel-Colman reaction

As originally described phthalimide is converted into the isocarbostyril on treatment with sodium ethoxide (S. Gabriel and J. Colman, Ber., 1900,33,2630). 3-Nitrophthalimide on reaction with methyl isocyanoacetate produces an oxazole which in turn is converted into the isoquinoline (K. Nunami *et al.*, Chem. Pharm. Bull., 1979, 27,1373; K. Nunami, M. Suzuki, and N. Yoneda, J. org. Chem., 1979, 44,1887).

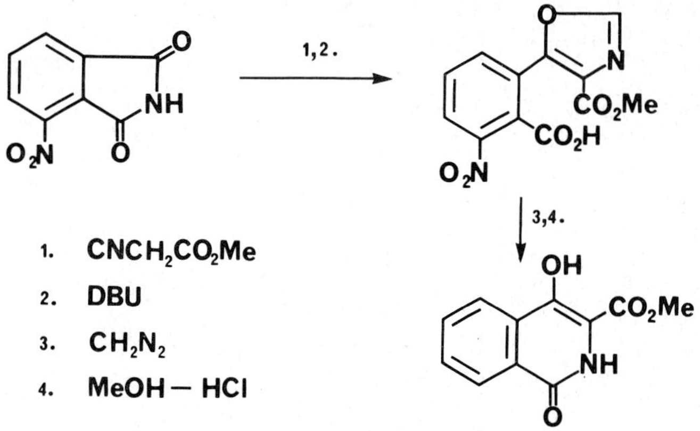

O
NH
O_2N
O
1,2.
O
N
CO_2Me
CO_2H
O_2N
3,4.
1. $CNCH_2CO_2Me$
2. DBU
3. CH_2N_2
4. MeOH — HCl
OH
CO_2Me
NH
O

The Gabriel-Colman procedure is not often used.

(iv)   Types 4 and 5 Syntheses

(1)   From homophthalates, isocoumarins and benzo[c]pyrylium salts

The formation of the N(2)-C(3) bond is involved when pre-
paring isoquinolines from homophthalic acid derivatives (R.
B. Tirodkar and R. N. Usgaonkar, Indian J. Chem., 1972, 10,
1060). Many analogues of these types of compounds in a
variety of oxidation states have been used to generate isoquinolines.

The ozonolysis of indenes followed by the action of ammonia
on the dialdehyde produced leads to isoquinolines in high
yields (R. B. Miller and J. M. Frincke, J. org. Chem.,
1980, 45, 5312).

R = H,1-Me,3-8 Me;5,8-Me$_2$,
6,7-(MeO)$_2$,6-7 Br, 6-7 I;
6-7 NO$_2$.

Yields = 57-96 %

Benzcyclobutenes have been used to generate *o*-formylbenzyl
ketones under flow vacuum thermolysis and these are converted into isoquinolines on treatment with ammonia (P.

Schiess, M. Huys-Francotte, and C. Vogel, Tetrahedron Letters, 1985, 26, 3959).

1. **550 – 600°C 0·01 mm Hg.**   **25 – 90 %**

Similarly, isocoumarins and benzo[c]pyrylium salts give isoquinolines on treatment with ammonia or amines (K. Ando, T. Tokoroyama, and T. Kubota, Bull. chem. Soc. Japan, 1974, 47, 1014). The latter has been extensively studied (E. V. Kuznetsov et al., Khim. Geterotsikl. Soedin., 1975, 25).

(2) *From 2-bromoaryl oximes*

The copper catalysed reaction of 2-bromoaldo- and keto-oximes with a variety of malonic acid derivatives results in the formation of 3-hydroxyisoquinoline-*N*-oxides in yields of 50-79% (A. McKillop and D. P. Rao, Synthesis, 1977, 760).

R²CH₂CO₂Et   1. NaH 2. CuBr.

(3) From 2-acyl-o-toluonitriles

The ketals derived from 2-aminomethylbenzyl ketones cyclize
readily when the ketone is deprotected in methanolic hydrogen chloride (C. K. Bradsher and T. G. Wallis, J. org. Chem.,
1978,43,3817).

(4) From 2-ethynylbenzaldehydes

The sulphonyl and acylhydrazones of 2-ethynylbenzaldehydes
give, in the presence of potassium carbonate or DBU, isoquinoline *N*-imines (P. N. Anderson and J. T. Sharp, J. chem.
Soc., Perkin I, 1980,1331).

(5) From 2-halogenobenzylamine derivatives

2-Halogenobenzylamines are convenient precursors of 1,2-
dihydroisoquinolines. Treatment with enolate anions leads
to intermediates which cyclize either spontaneously or on
treatment with Pd≐C. Treatment with reducing agents yields
the 1,2,3,4-tetrahydroisoquinolines (R. Beuglemans, J.
Chastanet, and G. Roussi, Tetrahedron, 1984, 40,311).

X = Br,Cl.

High yields of 3,4-dihydroisoquinolines are obtained by the lithiation of 2-halogenophenethyl halides and subsequent quenching with benzonitrile (C. A. Hergrueter *et al.*, Tetrahedron Letters, 1977,4145).

X = halogen

(6) *From vinyl azides*

The thermolysis of vinyl azides derived from ethyl azidoacetate and 2-methylbenzaldehydes gives isoquinolines in 50% yields on heating at reflux in toluene (T. L. Gilchrist, C. W. Rees, and J. A. R. Rodrigues, Chem. Comm., 1979,627; L. Henn *et al.*, J. chem. Soc., Perkin I, 1984,2189).

(7) From phthalides

Addition of the phthalide anion to a Schiff's base gives a product which subsequently ring closes to a 2,3-diaryl-isoquinoline. The latter undergoes rearrangement on treatment with trifluoroacetic acid to 2,4-diaryl-$1(2H)$-isoquinolones (D.J. Dodsworth et $al.$, Tetrahedron Letters, 1980,$\underline{21}$,5075).

(8) Dimerization of benzonitriles

1-Aminoisoquinolines can be produced by the dimerization of 2-methylbenzonitriles. The reaction proceeds in liquid ammonia/potassium amide and its utility is improved by the fact that two different nitriles can be used. Thus 1-amino-3-phenylisoquinoline is produced from 2-methylbenzonitrile and benzonitrile (H. Van der Goot and W. Th. Nauta, Chim. Therap., 1972,$\underline{7}$,185).

1. KNH_2 / NH_3.
2. PhCN
3. H_2O

(v) Type 6 Syntheses. The benzenoid ring

This approach to isoquinoline synthesis has received increasing attention. The need for isoquinolines saturated in the benzene ring (e.g. 5,6,7,8-tetrahydroisoquinoline) has contributed to this interest along with the increasing availability of many complex pyridine derivatives.

(1) Robinson annelation

The enamine derived from N-methyl-4-piperidinone and pyrrolidine on reaction with 3-oxopent-4-enoates generates the 1,2,3,4,7,8- octahydroisoquinoline (23) (G. Stork and R. N. Guthikonda, J. Amer. chem. Soc., 1972,94,5109).

(23)

Grignard reagents have been used to achieve similar conversions (J. A. Finkelstein and C. D. Perchonock, Tetrahedron Letters, 1980, 21,3323; M. A. Tius, A. Thurkauf, and J. W. Truesdell, Tetrahedron Letters, 1982,23,2819).

(2) Cycloaddition reactions

4-Vinylpyridine undergoes the Diels-Alder reaction with *N*-methylmaleimide to give a 7,8-dihydroisoquinoline. However, the reaction continues and the adduct (24) is isolated as the final product (T. Wagner - Jauregg, Q. Ahmed, and E. Pretsch, Helv., 1973,56,440).

(24)

4-Substituted-*N*-methyl-2-pyridones undergo the Diels-Alder reaction with 2,3-dimethylbutadiene leading to the fully aromatic isoquinoline after 96h at 190°C. At lower temperatures 1,2,5,8-tetrahydroisoquinolines are formed (H. Kato, R. Fujita, H. Hongo, and H. Tomisawa, Heterocycles, 1979,12,1).

R = CN , CO$_2$Et ; yield = 71,85%

1,2-Dihydropyridines undergo a Diels-Alder reaction with acrolein. Treatment of the adduct with a Wittig reagent followed by heating induces a Cope rearrangement to the tetrahydroisoquinoline (B. E. Evans *et al.*, J. org. Chem., 1979,44,3127; M. E. Christy *et al.*, *ibid.*, 1979, 44,3117).

The intramolecular Diels-Alder reaction of substitued aza-trienes gives a mixture of *cis-* and *trans-* octahydroiso-quinolines (S. F. Martin *et al.*, J. org. Chem., 1983,48, 5170). The reactions proceed at 25-275°C in 69-83% yield.

R = H, Me, CO₂Me; X = Y = H₂O.

Pyridynes generated by strong base (*e.g.* *n*-BuLi) from 3,4-dihalopyridines are trapped by furan. Treatment of these adducts with acid generates isoquinolines (D. J. Berry, B. J. Wakefield, and J. D. Cook, J. chem. Soc. (C), 1971,1227).

1-(4-Pyridyl)buta-1,3-diene on heating at 650°C produces 7,8-dihydroisoquinoline. The 3-pyridyl analogue yields mixtures of dihydroquinoline and dihydroisoquinolines (B. I. Rosen and W. P. Weber, J. org. Chem., 1977,42,47).

(vi) Type 7 Syntheses: Rearrangements

Flash vacuum thermolyses (FVT) of a number of trisubstitued 1,2,3-triazoles have been recorded.   At temperatures of 600 C these lead to a mixture of products which include substituted isoquinolines (T. L. Gilchrist, G. E. Gymer, and C. W. Rees, Chem. Comm., 1973,835).   Both 1,4-dimethyl-5-phenyl- and 1,5-dimethyl-4-phenyltriazole give 3-methyl isoquinoline suggesting a common intermediate the structure of which has been proposed as 1,2-dimethyl-3-phenylazirine (T. L. Gilchrist, G. E. Gymer, and C. W. Rees, J. Chem. Soc., Perkin I, 1973,555).

The pyrolytic transformation of 1,4-diacetoxy-2,3-diazidona-phthalene into an isoquinoline is achieved by briefly heating it in 1,2-dichlorobenzene. The major product is 3-cyanoisoquinoline (39%) which is believed to be derived by nitrene insertion, rearrangement of the azepine carbene and loss of nitrogen (Scheme 1) (D. S. Pearce, M.-S. Lee, and H. M. Moore, J. org. Chem., 1974,__39__,1362).

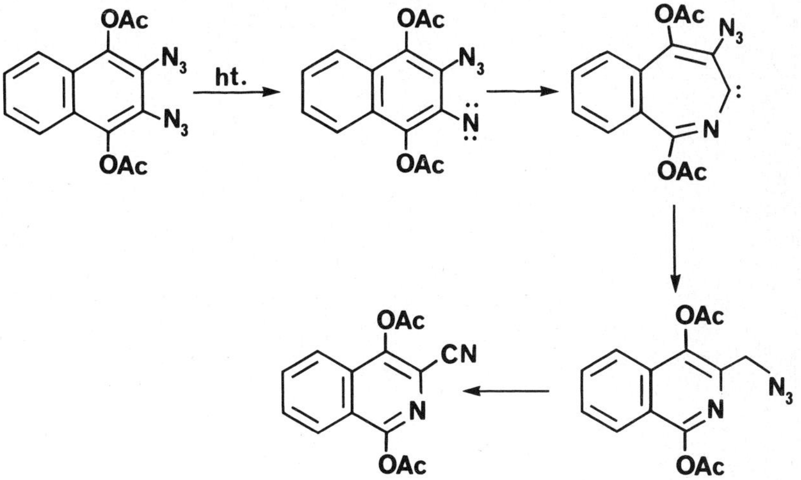

Scheme 1

Irradiation of certain isoindoles in dichloromethane in the presence of oxygen gives in addition to the corresponding anil a 42% yield of the phenanthridine (M. Ahmed, L. J. Kricka, and J. M. Vernon, J. chem. Soc., Perkin I, 1975,71).

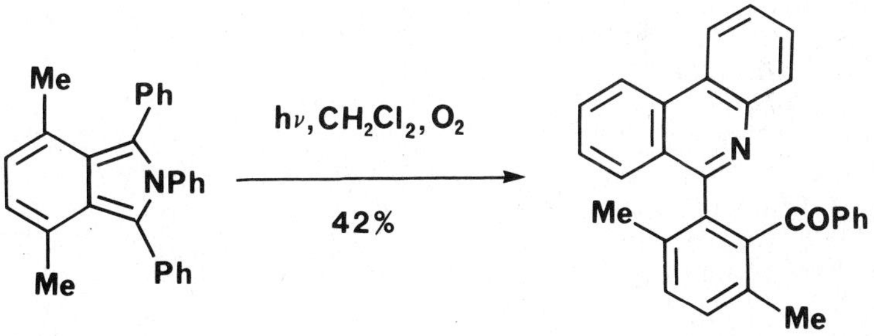

(b)   Physical properties of the isoquinolines

(i)   Proton and deuterium nmr spectra

Proton nmr spectra were discussed in the 2nd edition. Some
important work on reduced isoquinolines, particularly alkaloids, has appeared since (J. L. Moniot and M. Shamma,
Heterocycles, 1978, 9, 145).   It is found that chemical
shift values for B ring methylene protons increase in the
order: amines < protopines < enamimes < amides < imines <
imides < pyridones < N-oxides < pyridinium salts.   Incorporation of deuterium into isoquinoline has been followed by
deuterium nmr spectroscopy (J.M.A. Al-Rawi, G. Q. Behnam,
and N. I. Taha, Org. mag. Res. 1981, 17, 204).

(ii) ^{13}C-nmr spectra

The ^{13}C chemical shifts for isoquinoline have been reported
(R.J. Pugmire *et al.*, J. Amer. chem. Soc., 1969, 91, 6381).
Both the ^{13}C chemical shifts and coupling constants have
been measured for isoquinoline in $CDCl_3$ (S. R. Johns and R.
I.   Willing, Austral. J. Chem., 1976, 29, 1617).   Their
values are C1 152.5(178); C3 143.1 (178); C4 120.3 (161); C5
126.4 (161); C6 130.2 (161); C7 127.1 (163); C8 127.5
(161); C8a 128.7; C4a 135.7.   The values of ^{13}C chemical
shifts of some heterocycles in aqueous solution can vary by
as much as 10 ppm (for C-3 in isoquinoline) over a range of
pH.   A plot of chemical shift against pH gives a sigmoid
curve from which the pKa of isoquinoline can be determined
(E. Breitmaier and K.-H. Spohn, Tetrahedron, 1973, 29,
1145).   Studies have been made of the effect of methyl (J.
Su, E. Siew, E. V. Brown, and S. L. Smith, Org. mag. Res.,
1977, 10, 122), amino (L. M. Ernst, Org. mag. Res., 1976, 8,
161), and various other substituents (A. van Veldhuizen,
M. van Dijk, and G. M. Sanders, Org. mag. Res., 1980, 13,
105) on the ^{13}C-nmr spectra of isoquinolines.

(iii) ^{15}*N-nmr spectra*

As part of a general study of the nmr of azines the ^{15}N-nmr shielding for N in isoquinolines has been determined and compared with calculated values (L. Stefaniak *et al.*, Org. mag. Res., 1984, 22, 201). The effect of ring annelation and methyl substituents are discussed (M. Witanowski *et al.*, ibid., 1980, 14, 305). ^{14}N-shieldings for isoquinoline -2-oxide and other *N*-oxides have been determined.

(iv) Mass spectra

Elimination of hydrogen cyanide from the isoquinoline molecular ions has been studied by ^{13}C-labelling (M. A. Baldwin, J. Gilmore, and M. N. Mruzek, Org. mass Spec., 1983, 18, 127). No skeletal rearrangement takes place before HCN loss. Two-thirds of the HCN arises by C-1 loss and the rest comes from C-3. Studies of the mass spectral fragmentations of *N*-methylisoquinolinium ions (R. Salsmans and G. van Binst, *ibid.*, 1974, 8, 357) and aminoisoquinolines (E. V. Brown and S. R. Mitchell, *ibid.*, 1972, 6, 943) have been made.

(c) The chemical properties of isoquinolines

(i) Reduction

Reduction of isoquinolines has been reviewed (S. F. Dyke and R. G. Kinsman, in "Isoquinolines", ed. G. Grethe, Wiley, NY, 1981, Pt. 1, p. 1.). Pyridine: borane in acetic acid (a mild reducing agent) reduces isoquinoline to 1,2,3,4-tetrahydroisoquinoline (48% yield) (Y. Kikugawa, K. Saito, and S. Yamada, Synthesis, 1978, 447). 5-Nitroisoquinoline may be reduced under appropriate conditions to the corresponding di- or tetrahydro derivative without affecting the nitro group (K. V. Rao and D. Jackman, J. heterocyclic Chem., 1973, 10, 213).

74%   65%

1.NaBH$_4$, 25% MeOH,H$_2$O; 2.NaBH$_4$, HOAc

Sodium cyanoborohydride, in water, has been employed success-fully for the conversion of *N*-(4-nitrobenzyl)isoquinolinium bromide into *N*-(4-nitrobenzyl)-1,2,3,4-tetrahydroisoquinoline in 76% yield (R. O. Hutchins and N. R. Natale, Synthesis, 1979, 281). 1,2,3,4-Tetra- and 1,2-dihydroisoquinoline derivati-ves have been prepared by selective reduction (LAH or diborane) of 2-carbamoyl-1-phenyl-1,4-dihydroisoquinoline-3(2*H*)-ones (E. Zara-Kaczian, G. Deak, and G. Toth, J. chem. Res. (S), 1984, 282).

Tetrahydroisoquinolines may be obtained from isoquinoline-*N*-boranes, which are themselves stable, isolable inter-mediates that can be further elaborated (D. J. Brooks *et al.*, J. org. Chem., 1984, 49, 130).

1. **BH$_3$.THF, -78°C**; 2. **NaH, DBAH**; 3. **ClCO$_2$Me**; 4. **H$^+$**

A method for the assay and methylation of 2-methyl-1,2-dihydroisoquinoline (formed by LAH or sodium borohydride reduction of *N*-methylisoquinolinium iodide) that utilizes methyl iodide has been described (W. J. Gensler and K. T. Shamasundar, J. org. Chem., 1975, 40, 123). The synthesis and isomerization of 1,2,3,4,5,8-hexahydroisoquinolines have been reported (T. A. Crabb *et al.*, J. chem. Soc., Perkin I, 1975, 58; 1975, 1465). Stable 1,2-dihydroisoquinolines have been made by isomerization of 5,6-dihydroisoquinolines with potassium amide-liquid ammonia (T. R. Kasturi and L. Krishnan, Tetrahedron Letters, 1985, 865). Sodium hydrogen telluride has been found to reduce *N*-methylisoquinolinium iodide to the 1,2-dihydro derivative under basic conditions, but at pH 6 *N*-methyltetrahydroisoquinoline is obtained (D. H. R. Barton, A. Fekih, and X. Lusinchi, Tetrahedron Letters, 1985, 26, 3693). This reagent will also reduce the *N*-oxide (25) (96% yield) and the nitrone (26) (80% yield) to the corresponding 1,2,3,4-tetrahydroisoquinoline (D. H. R. Barton, A. Fekih, and X. Lusinchi, Tetrahedron Letters, 1985, 26, 4603).

$$(25) \qquad\qquad (26)$$

Isoquinolines normally undergo catalytic hydrogenation pre-
ferentially in the pyridine ring (P. Rylander in "Catalytic
Hydrogenation in Organic Synthesis", Academic, N.Y., 1979, p.
233). However, in strong acids high yield preferential hydrogenation of the benzene ring can be achieved (J. Z. Ginos, J.
org. Chem., 1975, 40, 1191; F. W. Vierhapper and E. L. Eliel,
ibid., 1975, 40, 2729).

Reduction of the hetero-ring, a nitro-group, and concurrent
hydrogenolysis of a bromo substituent can be achieved in a
"one-pot" process when a mixture of catalysts is employed
(I. W. Mathison and P. H. Morgan, J. org. Chem., 1974, 39,
3210).

1. HOAc (200ml); NH_4OAc $(25g)$; PtO_2,
Pd on $CaCO_3$, H_2 (40psi).

Isoquinoline undergoes hydrogenation to 1-formyl-1,2,3,4-
tetrahydroisoquinoline by iron pentacarbonyl (T. J. Lynch
et al., J. org. Chem., 1985, 49, 1266).

(ii) Oxidation

4-Hydroxy-5,6,7-trimethoxy-1,2,3,4-tetrahydroisoquinoline,
prepared by Bobbitt reaction, gives the quinolinol (27),
in reasonable yield, when heated at 140-145° with palladium-
carbon in *p*-cymene (M. P. Cava and I. Noguchi, J. org.
Chem., 1973, 38, 60).

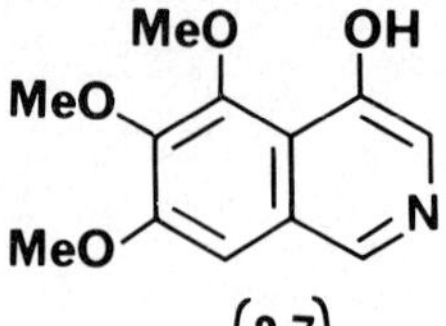

MeO, OH

MeO,

MeO

N

(27)

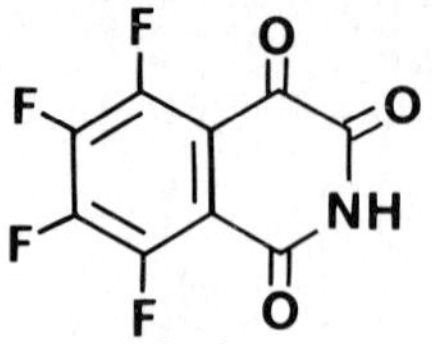

F O

F O

F NH

F O

(30)

Oxidation of 1,2,3,4-tetrahydroisoquinoline with Fremy's salt in sodium carbonate buffer proceeds stepwise *via* 3,4-dihydroisoquinoline, to isoquinoline (P. A. Wehrli and B. Schaer, Synthesis, 1974, 288).

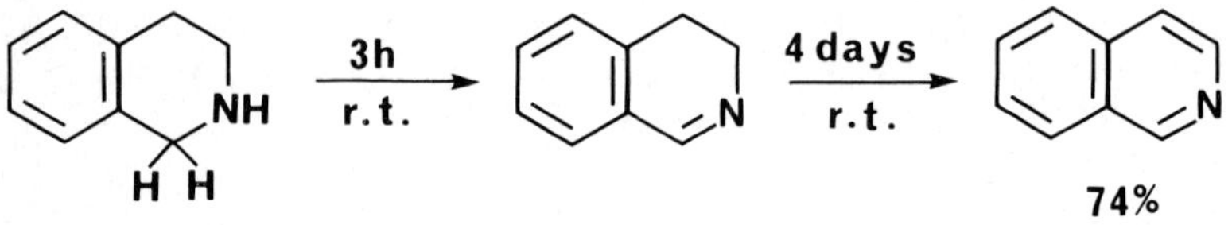

NH
H H

→ 3h r.t. →

=N

→ 4 days r.t. →

N

74%

N-Hydroxy-1,2,3,4-tetrahydroisoquinoline is readily converted into 3,4-dihydroisoquinoline in high yield on treatment with titanium (III)chloride (S. Murahashi and Y. Kodera, Tetrahedron Letters, 1985, 26, 4633). The 7-and 8-quinolinols undergo copper-catalyzed oxidation, with oxygen in the presence of morpholine in good yield (H. Fukumi, H. Kurihara, and H. Mishima, J. heterocyclic Chem., 1978, 15, 569; Y. U. Tsizin, Khim. Geterotsikl. Soedin., 1974, 1253). Further treatment of (28) with methanol and sulphuric acid yields (29).

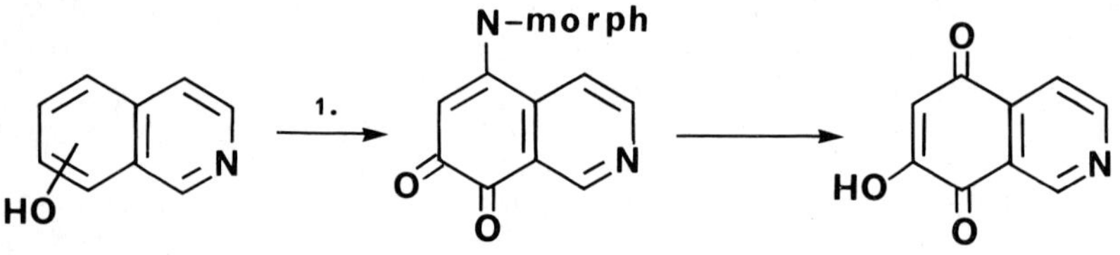

N-morph

O

HO

N

→ 1. →

O

O

N

→

HO

O

O

N

(28)   (29)

1. O_2, $Cu(OAc)_2$, morpholine.

Oxidation across the 3,4-double bond in some 1,2-dihydro isoquinolines is readily achieved by chromium (VI) oxide in acetic acid (J. Urbanski and L. Wrobel, Pol. J. Chem., 1984, 58, 899).

R = COPh, SO₂Ph

Perfluoroisoquinoline gives the trioxo-derivative (30) on reaction with 98% nitric acid at 100°C (P. Sartori, K. Ahlers, and H.-J. Frohn, J. fluorine Chem., 1976, 457).

1,2,3,4-Tetrahydroisoquinoline undergoes sodium tungstate catalyzed oxidation with hydrogen peroxide to afford the nitrone (26) in 85% yield (H. Mitsui, S. Zenki, T. Shiota, and S. Murahashi, Chem. Comm., 1984, 874). Intramolecular oxidative coupling of isoquinolines has been reviewed (O.P. Dhingra in "Oxidation in Organic Chemistry", ed., W. S. Trahanovsky, Academic, 1982, Pt.D, p. 207).

(iii) Substitution reactions

Isoquinolines undergo nucleophilic attack most readily at C-1, but reaction may occur at C-3. 1-Methylisoquinoline (72-76% yield) has been prepared by ultrasound irradiation of a solution of isoquinoline and sodium hydroxide in DMSO for 2h at room temperature (J. Ezquerra and J. Alvarez-Builla, Org. prep. Proceed. Int., 1985,17,190). Isoquin-olines are aminated in high yield at position 1 under Chi-chibabin reaction conditions and the kinetics of this reaction have been studied (V. N. Novikov, A. F. Pozharskii, and V. N. Doron'kin, Khim. Geterosikl. Soedin. 1976, 244).

The positional order of electrophilic reactivity of the neutral isoquinoline molecule has been found by pyrolysis of 1-arylethyl acetates to be 4 > phenyl > 5 = 7 > 8 > 6 > 3 > 1 (E. Glyde and R. Taylor, J. chem. Soc., Perkin II, 1975,

1783). Therefore, a special mechanism to account for 4-bromination of isoquinoline as its free base is unnecessary. Treatment of a slurry of isoquinolinium hydrochloride in nitrobenzene gives 4-bromoisoquinoline in 84% yield (T. J. Kress and S. M. Constantino, J. heterocyclic Chem., 1973, 10, 409). Isoquinoline and its N-oxide both undergo nitration (conc. nitric and conc. sulphuric acids) as their conjugate acids at the 5- and 8- positions (J. Gleghorn _et al_., J. chem. Soc. (B), 1966, 870). However, 1-cyanoisoquinoline-2-oxide gives mainly 6-nitroisoquinoline-2-oxide (58%) on nitration with fuming nitric acid (M. Hamana and H. Saito, Heterocycles, 1977, 8, 403).

Acid-catalyzed hydrogen-deuterium isotope exchange takes place at positions C-5 and C-8 of the isoquinolinium ion between D_0 -5 to -8.5 at 180°C (U. Bressel, A. R. Katritzky, and J. R. Lea, J. chem. Soc. (B), 1971,4).

A study of the free radical phenylation of isoquinoline has revealed a reactivity order of 1 > 5 > 8 > 4 > 3, 6, 7 (L. K. Dyall and C. J. Pullin, Austral. J. Chem., 1979, 32, 345).

(iv) Isoquinolinium salts

Two principal reactions of isoquinolinium salts are dealt with in this section, addition-elimination (see also (vii) Reissert Compounds) and 1,2-addition followed by ring-opening. Typical reactions of pyridinium, quinolinium, and isoquinolinium salts have been compared (E. F. V. Scriven in "Comprehensive Heterocyclic Chemistry", ed. A. R. Katritzky and C. W. Rees, Pergamon, 1984, Vol. 2).

Equilibrium constants for pseudo base formation from isoquinolinium cations have been measured (J. W. Bunting and D. J. Norris, J. Amer. chem. Soc., 1977, 99, 1189) and substituent effects correlated by Hammett equations (J. W. Bunting, V. S. F. Chew, and S. Sindhuatmadja, Canad. J. Chem., 1981, 59, 3195). The kinetics of oxidation of isoquinolinium cations with ferricyanide ions have been studied (J. W. Bunting, P. A. Lee-Young, and D. J. Norris, J. org. Chem., 1985, 43, 1132). Covalent amination of isoquinolinium cations has been studied by nmr (J. A. Zoltewicz and J. K. O'Halloran, J. Amer. chem. Soc., 1975, 97, 5531). Ring-opening of the isoquinolinium system and subsequent reclosure to form a naphthalene is an old reaction (R. Decker, Ann., 1908, 362, 305). However, modification of the reaction conditions has led to a useful synthetic approach to 1-naphthylamines (A. N. Kost *et al.*, Chem. heterocyclic Compounds, 1980,1117).

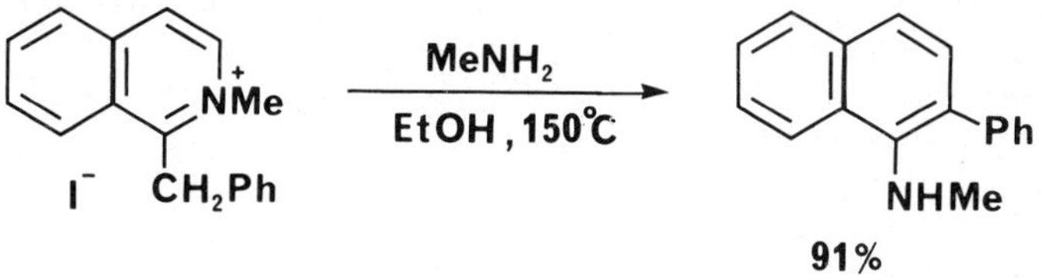

91%

Alkylation of the C-1 position in isoquinolines is a key reaction in the synthesis of alkaloid precursors. Traditionally anions of Reissert compounds have been used to achieve such a conversion. Now the reactions of boron enolates or trimethylsilyl enol ethers provide a viable alternative (K. Akiba *et al.*, Tetrahedron Letters, 1981, 22, 4961; J. org. Chem., 1985, 50, 63).

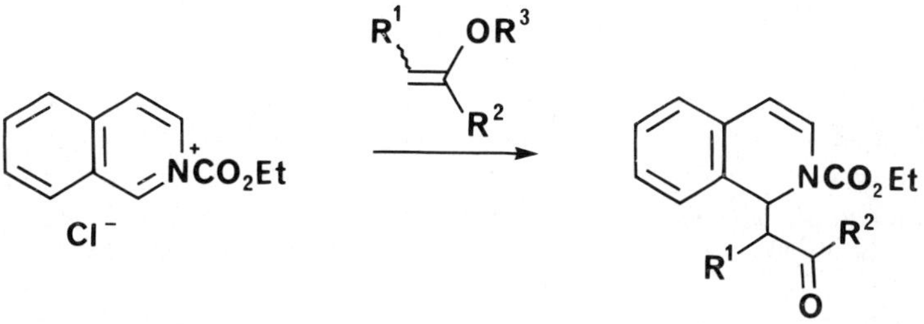

Isoquinoline forms adduct (31) on treatment with *N*-methylpyrrole and a sulphonyl chloride, and adduct (32) with ferrocene and an imidoyl chloride (A.K. Sheinkman, L. M. Sidorenko, and Y. G. Skrypnik, Chem. heterocyclic Compounds, 1985, 1136; and I. Y. Kozak, G. N. Yashchenko, and A. K. Sheinkman, *ibid.*, 1985,814).

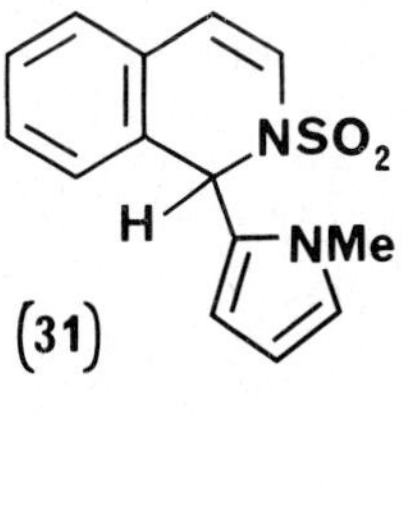

(31)

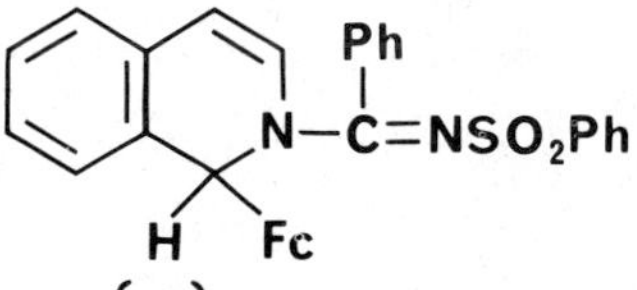

(32)

Isoquinoline also forms C-1 coupled products with methylbutenolides (R. J. Rafka and W. A. Szarek, Heterocycles, 1984, 22, 2019).

Isoquinoline 2-oxides readily undergo nucleophilic addition at C-1 in the presence of acylating agents, subsequent elimination takes place to give for example (33) and (34) (M.M. Yousif, S. Saeki, and M. Hamana, Chem. Pharm. Bull., 1982, 30, 1680; K. Funakoshi, H. Inada, and M. Hamana, *ibid.*, 1984, 32, 4731).

(33)

(34)

(v) Reactions involving substituents

Reaction of 6,7-dimethoxy-1-lithiomethylisoquinoline with electrophiles affords useful isoquinoline building blocks for alkaloid synthesis (35) (E. M. Kaiser and P. L. Knutson, Synthesis, 1985, 148).

(35)

(36)

3,4-Isoquinolinedione-4-oxime hydrate (36) has been reported to be the previously unidentified by-product of the reaction of 3-aminoisoquinoline with nitrous acid (T. J. Schwan and H. A. Burch, J. heterocyclic Chem., 1983, 20, 239). 1-Benzyl-3,4-dihydroisoquinoline is oxidized regiospecifically by singlet oxygen (N. H. Martin, S. L. Champion, and P. B. Belt, Tetrahedron Letters, 1980, 21, 2613).

99%

1-Substituted-3-methoxy-4-cyano-5,6,7,8-tetrahydroisoquinol-
ines undergo chlorination (at C-3) in good yield under
Vilsmeier-Haack conditions. However, if the 1-substituent
is a strong resonance donor chlorination is not favoured (T.
R. Kasturi, H. R. Y. Jois, and L. Mathew, Synthesis, 1984,
743).

(37)

Treatment of *N*-tosylamino-1,2,3,4-tetrahydroisoquinoline
with base gives the dimer (37) in almost quantitative yield
(B. F. Powell, C. G. Overberger, and J. P. Anselme, J.
heterocyclic Chem. 1983, 20, 121). The 3-azidoisoquinoline-
tetrazoloisoquinoline tautomerism has been studied; rather
surprisingly the tetrazole tautomer is favoured in TFA solution (A. Messmers and G. Hajos, J. org. Chem., 1981, 46,
843).

(vi)   Rearrangement reactions

Photolysis of 4- and 5-isoquinolyl azides in primary aliphatic amines affords diazepines (38) or azepines (39) (F.
Hollywood *et al.*, J. chem. Soc., Perkin I, 1982, 421).
Similar ring expansions are observed on photolysis of the
same azides in methanol containing sodium methoxide (F.
Hollywood *et al.*, J. chem. Soc., Perkin I, 1982, 431),
however photolysis of 4-azidoisoquinoline and its *N*-oxide
in hydrohalogenic acids gives 3-halogeno-4-amino derivatives
(H. Sawanishi, T. Hirai, and T. Tsuchiya, Heterocycles,
1982, 19, 1043).

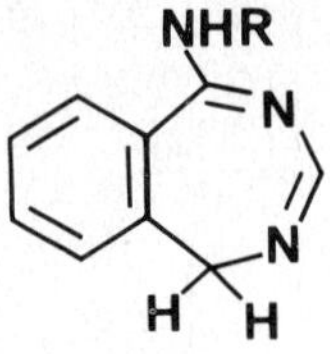

(38) 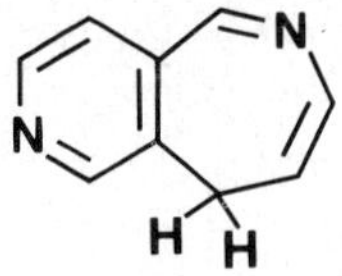 (39) 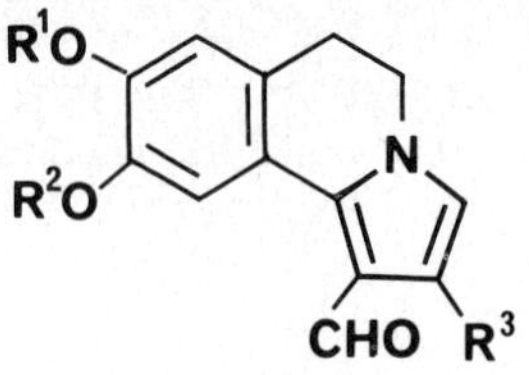 (41)

Irradiation of 1-methoxyisoquinoline 2-oxide in benzene yields the benzoxazepine (40) (A. Albini, E. Fasani, and L. M. Dacrema, J. chem. Soc., Perkin I, 1980, 2738).

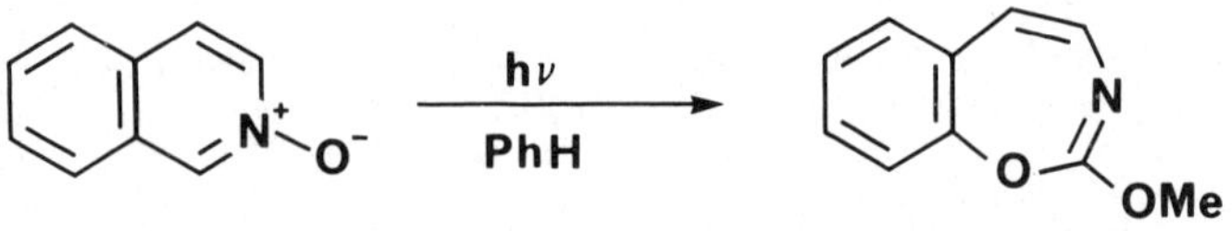

(40) 40%

Photolysis of isoquinoline N-imides gives $1H$-1,3-benzo-diazepines (T. Tsuchiya et $al.$, Chem. Comm., 1979, 534). This method offers a route to diazepines that are not available by azide ring-expansion. Another ring-expansion sequence involves the addition of dichlorocarbene to isocar-bostyrils (U. K. Pandit, Symp. Heterocycl., 1977, 22). Cycloprop c isoquinolines undergo two types of thermal rearrangement to $1H$ (or $5H$) 2-benzazepines depending upon the nature of the substituents (K. Motion, I. R. Robertson, and J. T. Sharp, Chem. Comm., 1984, 1531).

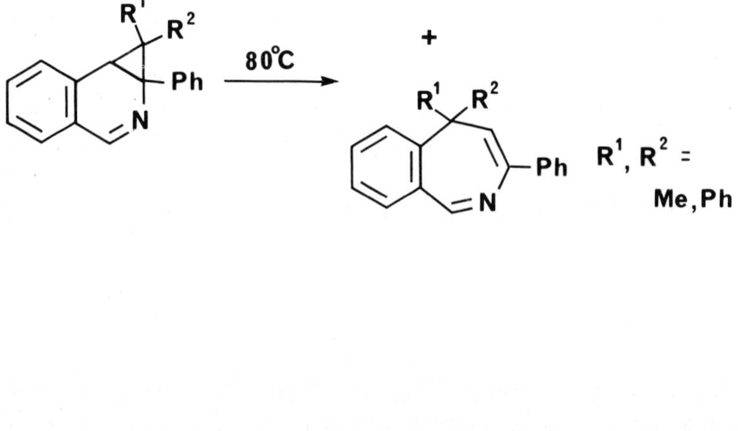

R^1 = H
R^2 = Me, Ph

R^1, R^2 =
Me, Ph

Amination of 1-bromoisoquinoline occurs by the addition-elimination mechanism, but amination of 3-bromoisoquinoline chiefly (55%) involves the S_n(ANRORC) pathway. The remaining 45% of substitution probably goes by the addition- elimination pathway (G. M. Sanders, M. van Dijk, and H. J. den Hertog, Rec. trav. Chim., 1974, 93, 198). 1-(3-Furanyl)-3,4-dihydroisoquinolines rearrange to pyrroloisoquinolines (41) on heating in a high boiling amine (W. Loesel and H. Daniel, Chem. Ber., 1985, 118, 413).

(vii) Reissert compounds

Interest in Reissert compounds derived from isoquinoline and other compounds abounds, and this is reflected in the number of reviews on the subject (F. D. Popp, Adv. heterocyclic chem., 1968, 9, 1; 1979, 24, 187; F. D. Popp, Heterocycles, 1973, 1, 165; 1980, 14, 1033; J. V. Cooney, J. heterocyclic Chem., 1983, 20, 823; E. F. V. Scriven in "Comprehensive Heterocyclic Chemistry", Pergamon, 1984, Vol. 2, p. 248; F. D. Popp and B. C. Uff, Heterocycles, 1985, 23, 731).

Several superior methods for the preparation of Reissert compounds have appeared. These overcome problems inherent in older methods that utilize potassium cyanide and an acid chloride in aqueous solution, where both starting materials and product are not soluble in water, the halide can hydrolyse, and pseudo base formation might interfere. Addition of a phase transfer reagent can result in a spectacular increase in yield of Reissert compound (B. C. Uff and R. S. Budhram, Heterocycles, 1977, 6, 1789).

| 1. H_2O, CH_2Cl_2 | 15% | 2% |
| 2. H_2O, CH_2Cl_2, $PhCH_2\overset{+}{N}Me_3$ | - | 72% |

Other examples of the use of alkylammonium salts as phase
transfer agents have been described (D. Bhattacharjee and F.
D. Popp, Heterocycles, 1977, 6, 1905; Y. Hamada, K.
Morishita, and M. Hirota, Chem. Pharm. Bull., 1978, 26,
350; T. Koizumi, K. Takeda, K. Yoshida, and E. Yoshii,
Synthesis, 1977, 497). Potassium cyanide and 18-crown-
6-ethers have been used (R. Chenevert, E. Lemieux, and N.
Voyer, Synth. Comm., 1983, 13, 1095). The use of trimethylsilyl cyanide (and a catalytic amount of AlCl3)
instead of potassium cyanide as the source of cyanide
obviates the difficulties encountered when working under
aqueous conditions and high yields have been obtained using
this reagent (S. Ruchirawat *et al.*, Heterocycles, 1977, 6,
43; S. Ruchirawat and P. Thepchumrune, Org. prep. Proced.
Int., 1980, 12, 263; D. Bhattacharjee and F. D. Popp, J.
heterocyclic Chem., 1980, 17, 1207, 1211; S. Veeraraghavan
and F. D. Popp, *ibid.*, 1981, 18, 71; S. Veeraraghavan, D.
Bhattacharjee, and F. D. Popp, *ibid.*, 1981, 18, 443).
Acetone cyanohydrin (N. L. Sergovskaya, Yu. S. Tsizin, and
S. A. Chernyak, Chem. heterocyclic Compds., 1984, 654) and
tri-*n*-butyltin cyanide (F. D. Popp and J. Kant, Heterocycles, 1985, 23, 2193) have also been employed as sources
of cyanide. The well-known instability of *N-p*-tosyl
Reissert compounds has been exploited for the synthesis of
1-cyanoisoquinolines (D. L. Boger *et al.*, J. org. Chem.,
1984, 49, 4056).

1. Ts Cl, KCN, CH₂Cl₂ ; 2. DBU ,THF, 25°C.

The elimination in the above sequence also may be achieved
by air oxidation (M. D. Rozwadowska and D. Brozda, Canad. J.
Chem., 1980, 58, 1239). Chlorocyanation has been carried
out using sulphuryl chloride as the acylating agent (G. W.
Kirby, S. L. Tan, and B. C. Uff, J. chem. Soc., Perkin I,
1979, 270).

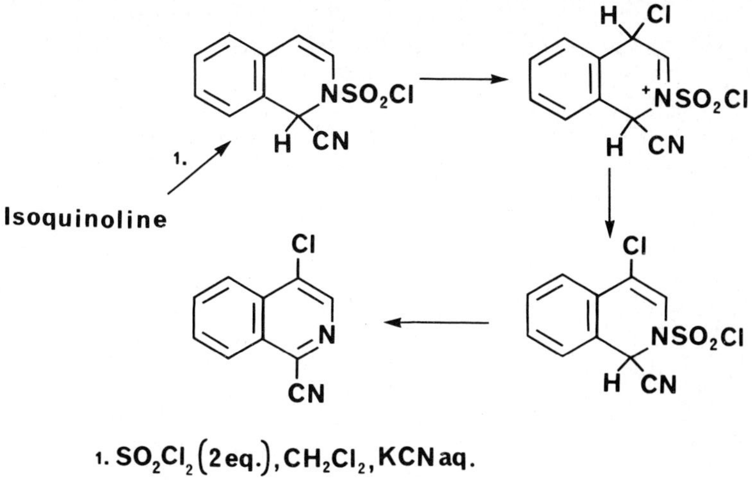

Isoquinoline

1. SO_2Cl_2 (2 eq.), CH_2Cl_2, KCN aq.

Alkylation of Reissert compounds is an area of intense activity because of the importance of this reaction for the synthesis of alkaloids of medicinal interest, *vid.* the reviews above. For instance, potential anticonvulsants (42) can be prepared in good yield by Reissert anion methodology (F. D. Popp and J. Kant, J. heterocyclic Chem., 1985, 22, 869). A comparative study of methods for the alkylation of Reissert compounds has appeared (J. W. Skiles and M. P. Cava, Heterocycles, 1978, 9, 653). Lithiated *N*-pivaloyl tetrahydroisoquinolines have been found to be super-nucleophiles that can be "titrated" with electrophiles, and therefore offer an advantage over traditional Reissert compounds as alkaloid precursors (J. - J. Lohman *et al.*, Angew. Chem., intern. Edn., 1981, 20, 128).

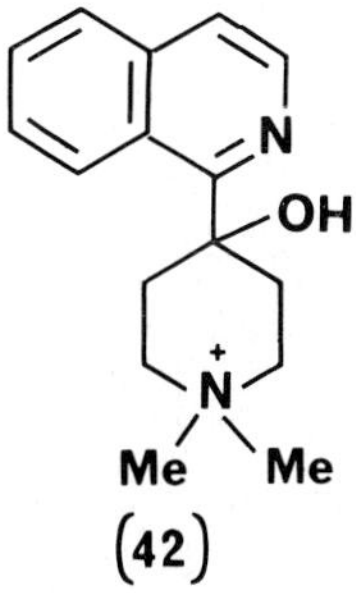

(42)

(viii)   Cycloadditions

Cycloadditions of isoquinoline salts with electron-rich olefins which take place to form tricyclic systems regiospecifically have been well studied (T. - K. Chen and C. K. Bradsher,
J. org. Chem., 1979, 44, 4680; C. K. Bradsher in "Isoquinolines",
ed. G. Grethe, Wiley, N.Y., 1981, Pt. 1, p. 381).   Addition
of enol ethers of cyclic ketones to isoquinolinium salts
yields anthracene derivatives (R. W. Franck and R. B. Gupta,
Tetrahedron Letters, 1985, 26, 293).

74%

A similar cycloaddition has been used in the synthesis of
epicorynoline (J. R. Falck and S. Manna, J. Amer. chem.
Soc., 1983, 103, 631).   Intramolecular cycloadditions also
have been studied (G. P. Gisby, P. G. Sammes, and R. A.
Watt, J. chem. Soc., Perkin I, 1982, 249).

2.   Hydroisoquinolines

(a)   1,2-Dihydroisoquinolines

Reduction of isoquinolines or their quaternary salts with
complex metal hydrides results in the formation of
1,2-dihydroisoquinoline.   Often these are converted further
to 1,2,3,4-tetrahydroisoquinolines particularly with borohydrides in protic solvents (J. G. Keay in Adv. heterocyclic
Chem., 39, 1).   Lithium aluminium hydride in diethyl ether
generates 1,2-dihydroisoquinolines.   For reduction with
sodium borohydride, isoquinolines normally require an
electron-withdrawing group in the 4-position (Y. Kikugawa,
Chem. Pharm. Bull., 1973, 21, 1914).

(b)   3,4-Dihydroisoquinolines

These are generated in the Bischler-Napieralski reaction as discussed under Type I syntheses.

(c)   1,2,3,4-Tetrahydroisoquinolines

Isoquinolines with a completely saturated pyridine ring may be prepared directly from benzene precursors or by reduction of 3,4-dihydroisoquinolines formed in the Bischler-Napieralski reaction as discussed above. 1,2,3,4-Tetrahydroisoquinolines are also available by both chemical and catalytic reduction of isoquinolines. Catalytically, copper chromite, Raney nickel and platinum may be used but these reactions are not always successful, with significant overreduction sometimes occurring. Chemical reduction is the preferred method for quaternary isoquinolinium salts. Sodium borohydride in a protic solvent is often used, as well as lithium aluminium hydride. In carboxylic acids, with borohydrides, isoquinolines undergo reductive alkylation (R. V. Rao and D. Jackman, J. heterocyclic Chem., 1973,10,213; G. W. Gribble and P. W. Heald, Synthesis, 1975,650).

(d)   5,6,7,8-Tetrahydroisoquinolines

Catalytic hydrogenation of isoquinoline to the completely saturated decahydroisoquinoline is followed by dehydrogenation with palladium black leading to mixtures of isoquinoline and 5,6,7,8-tetrahydroisoquinoline (S. F. Dyke and P. G. Kinsman in "Isoquinolines" ed. G. Grethe, The Chemistry of Heterocyclic Compounds eds., A. Weissberger and E. C. Taylor, John Wiley, Pt. 1, 113, 1981).

Careful hydrogenation of isoquinoline can be employed to yield the tetrahydroisoquinoline directly (F. W. Vierhapper and E. L. Eliel, J. Amer. chem. Soc., 1974,96,2256).

95%

Guide to the Index

This index is constructed in a similar manner to the volume indexes of the first edition of the Chemistry of Carbon Compounds. However, to make the index easier to use, more descriptive entries have been made for the commonly occurring individual, and groups of chemicals.

The indexes cover primarily the chemical compounds mentioned in the text, and also include reactions and techniques, where named, and some sources of chemical compounds such as plant and animal species, oils, etc.

Chemical compounds have been indexed alphabetically under the names used by authors, editing being restricted to ensuring uniformity of entries under the same heading. In view of the alternative nomenclature that can often be used, a limited amount of cross-referencing has been done where it is considered to be helpful, but attention is particularly drawn to Convention 2 below.

For this and the succeeding volumes, the indexing conventions listed below have been adopted.

1. *Alphabetisation*

(a) The following prefixes have not been counted for alphabetising:

n-	*o-*	*as-*	*meso-*	D	C
sec-	*m-*	*sym-*	*cis-*	DL	*O-*
tert-	*p-*	*gem-*	*trans-*	L	*N-*
	vic-				*S-*
		lin-			*Bz-*
					Py-

Some prefixes and numbering have been omitted in the index, where they do not usefully contribute to the reference.

(b) The following prefixes have been alphabetised:

Allo	Epi	Neo
Anti	Hetero	Nor
Cyclo	Homo	Pseudo
	Iso	

(c) A letter by letter alphabetical sequence is followed for entries, firstly for the main entry, followed by the descriptive entry. The only exception to this sequence is the placing of plural entries in front of the corresponding individual entries to prevent these being overlooked by a strict alphabetical sequence which could lead to a considerable separation of plural from individual entries. Thus "butanes" will come before n-butane, "butenes" before 1-butene, and 2-butene, etc.

2. Cross references

In view of the many alternative trivial and systematic names for chemical compounds, the indexes should be searched under any alternative names which may be indicated in the main body of the text. Only a limited amount of cross-referencing has been carried out, where it is considered that it would be helpful to the user.

3. Esters

In the case of lower alcohols esters are indexed only under the acid, e.g. propionic methyl ester, not methyl propionate. Ethyl is normally omitted e.g. acetic ester.

4. Derivatives

Simple derivatives are not normally indexed if they follow in the same short section of the text.

5. Collective and plural entries

In place of "– derivatives" or "– compounds" the plural entry has normally been used. Plural entries have occasionally been used where compiunds of the same name but differing numbering appear in the same section of the text.

6. Main entries

The main entry of the more common individual compounds is indicated by heavy type. Multiple entries, such as headings and sub-headings over several pages are shown by "–", e.g., 67–74, 137–139, etc.

INDEX

250

DATE DUE